AF745166

# Electromagnetic Field - From Atomic Level to Engineering Applications

*Edited by Chao Wang and Yifan Kang*

Published in London, United Kingdom

Electromagnetic Field - From Atomic Level to Engineering Applications
http://dx.doi.org/10.5772/intechopen.1005982
Edited by Chao Wang and Yifan Kang

Contributors
Aishwarya Pradeep, Akshita Yadav, Andrea Bailey, Anthony Bassesuka Sandoka Nzao, Arbab I. Arbab, Chao Wang, Gennady Morozov, Himangshu B. Baskey, Hongtao Yang, Javier Resta-López, Jolanta M. Wołowicz, Joseph R. Salvatore, Natalya Stakhova, Oleg Morozov, Pablo Martín-Luna, Prashant S. Alegaonkar, Priyanka, Ryszard Wojnar, Somdotta Roy Choudhury, Sujatha Basa Reddy, Venkatagiri Bharath, Vladimir Anfinogentov, Wei Zhao, Yifan Kang, Yuri Sedelnikov

© The Editor(s) and the Author(s) 2025

The rights of the editor(s) and the author(s) have been asserted in accordance with the Copyright, Designs and Patents Act 1988. All rights to the book as a whole are reserved by INTECHOPEN LIMITED. The book as a whole (compilation) cannot be reproduced, distributed or used for commercial or non-commercial purposes without INTECHOPEN LIMITED's written permission. Enquiries concerning the use of the book should be directed to INTECHOPEN LIMITED rights and permissions department (permissions@intechopen.com)

Violations are liable to prosecution under the governing Copyright Law.

Individual chapters of this publication are distributed under the terms of the Creative Commons Attribution 4.0 License which permits commercial use, distribution and reproduction of the individual chapters, provided the original author(s) and source publication are appropriately acknowledged. If so indicated, certain images may not be included under the Creative Commons license. In such cases users will need to obtain permission from the license holder to reproduce the material. More details and guidelines concerning content reuse and adaptation can be found at http://www.intechopen.com/copyright-policy.html.

Notice

Statements and opinions expressed in the chapters are these of the individual contributors and not necessarily those of the editors or publisher. No responsibility is accepted for the accuracy of information contained in the published chapters. The publisher assumes no responsibility for any damage or injury to persons or property arising out of the use of any materials, instructions, methods or ideas contained in the book.

First published in London, United Kingdom, 2025 by IntechOpen
IntechOpen is the global imprint of INTECHOPEN LIMITED, registered in England and Wales, registration number: 11086078, 167-169 Great Portland Street, London, W1W 5PF, United Kingdom

For EU product safety concerns: IN TECH d.o.o., Prolaz Marije Krucifikse Kozulić 3, 51000 Rijeka, Croatia, info@intechopen.com or visit our website at intechopen.com.

British Library Cataloguing-in-Publication Data
A catalogue record for this book is available from the British Library

Electromagnetic Field - From Atomic Level to Engineering Applications
Edited by Chao Wang and Yifan Kang
p. cm.
Print ISBN 978-1-83634-725-5
Online ISBN 978-1-83634-724-8
eBook (PDF) ISBN 978-1-83634-726-2

If disposing of this product, please recycle the paper responsibly.

IntechOpen
intechopen.com

Built by scientists, for scientists

Explore all IntechOpen books

# Meet the editors

Dr. Prof. Chao Wang received his Ph.D. degree in physical electronics in 2011 from Xi'an Jiaotong University, China. Since then, he has been a full-time researcher at the Xi'an Institute of Optics and Precision Mechanics, a part of the Chinese Academy of Sciences. He visited the Department of Physics at Kansas State University, USA, from 2008 to 2010, and the Max-Planck Center for Attosecond Science at POSTECH, South Korea, from 2014 to 2015. He has been engaged in research on the frontier cross-application of attosecond laser technology and metamaterial technology. As the project chief designer, he has presided over more than ten national major research projects, authored three monographs and one translation, and written over 50 articles and 20 patents. He serves as the long-term reviewer of several international academic journals.

Prof. Yifan Kang received her bachelor's and master's degrees in optical materials from Northwestern University, China, in 2003 and 2008, respectively. Since then, she has been a full-time, research-oriented teacher at the Air Force Engineering University in China and has won the Excellent Teacher Award multiple times. Her research interests include ultrafast diagnostics, nanophotonics and metamaterials. To date, she has authored more than thirty publications and two monographs.

# Contents

# Preface

Electromagnetic fields, as one of the fundamental theories for understanding the workings of the material world, have persistently remained at the core of modern technological innovation. Whether it is the wireless communications and high-speed computing chips that underpin the foundation of the Internet of Everything, the diverse sensors and new energy equipment that drive intelligent control operations, or the astonishingly precise detection capabilities behind medical imaging instruments, electromagnetic field technology permeates them all, serving as a silent yet fundamental driving force. However, conventional perspectives often artificially separate macroscopic engineering techniques from their inherent microscopic quantum essence. This limits our deeper insight into the connections between electromagnetic phenomena and materials/devices, and also hinders our ability to precisely control, optimize, and even reshape these fields at higher dimensions. Achieving a proper integration from microscopic foundations to engineering top levels has become a critical pathway for breaking through existing technological bottlenecks and unleashing novel productive capacities.

This book is dedicated to constructing a bridge integrating theory and engineering across this divide. We will first look into the interiors of atoms and molecules, scrutinizing within the quantum world how charges and spins manifest new orders and give rise to novel effects, such as surface plasmon resonance, magnetic anisotropy manipulation, and artificial microstructures of metamaterials, under the influence of precisely configured electromagnetic fields. We will thoroughly examine the underlying physical principles and mathematical language that give rise to these effects. Subsequently, we reveal how material response parameters (permittivity, permeability, conductivity, etc.), inherently embedding these microscopic mechanisms, are integrated into macroscopic electromagnetic equations. This elucidates the complete journey of electromagnetic waves, from generation and control to propagation, and the intrinsic laws governing how the properties of media constrain them. Finally, focusing on core technological applications in representative scenarios such as electromagnetic invisibility/cloaking, electromagnetic shielding, electromagnetic compatibility (EMC), electromagnetic medical applications, and electromagnetic propulsion, we progressively weave the profound language of atomic physics into a powerful and precise blueprint for practical engineering.

As the Academic Editors, we would like to express our deepest respect and heartfelt gratitude to all the scholars who have contributed to this monograph. With profound expertise and innovative perspectives, the chapter authors have systematically constructed a comprehensive knowledge system spanning from quantum foundations to engineering applications. Each chapter embodies years of dedicated intellectual effort, from groundbreaking interpretations of microscopic mechanisms to precise derivations of macroscopic designs and breakthrough solutions for industry pain points. Every page resonates with scholarly dedication and technological foresight.

This monumental work will undoubtedly become an essential reference for researchers and a groundbreaking guide for engineers. The editorial team is profoundly honored to have witnessed the convergence of such brilliant insights. Sincere thanks also go out to the Publishing Process Manager, Elvira Baumgartner, and all staff members at IntechOpen. Their unwavering professionalism and invaluable contributions deserve the deepest appreciation.

**Chao Wang**
Xi'an Institute of Optics and Precision Mechanics of Chinese Academy of Sciences,
Xi'an, China

**Yifan Kang**
Air Force Engineering University,
Xi'an, China

Section 1

# New Mechanisms of Atomic Level Electromagnetic Field Effects

Chapter 1

# Quantum Electrodynamics of Massive Bosons

*Arbab I. Arbab*


## Abstract

This chapter presents a novel theoretical framework that extends quantum electrodynamics to incorporate massive bosons using a quaternionic formalism. By modifying Maxwell's equations to include a mass term, the study reveals that massive bosons exhibit unique electromagnetic properties: They can sustain a nonzero electric field even when the magnetic field vanishes under suitable gauge transformations. Unlike traditional electrodynamics, which strictly follows the Lorenz gauge condition, this formulation highlights the essential role of boson mass in defining quantum energy and momentum densities. It naturally leads to field equations that satisfy the Klein-Gordon equation and demonstrates a deep equivalence between particle and wave momentum, reinforcing wave-particle duality. The framework predicts novel effects, such as classical Hall conductivity without an external magnetic field and a quantized Hall effect in two-dimensional electron gases (2DEGs). It suggests that phenomena like quantum inductance, critical currents, and quantized capacitance can be directly linked to the mass and spin of bosons. Additionally, the theory introduces gauge-spin transformations that couple spin and helicity densities to the boson's mass, offering new perspectives on energy-momentum conservation in electromagnetic systems. The model aligns with the established theories, including Wilczek's axion electrodynamics and the Proca formulation, while opening new experimental possibilities in condensed matter physics, particularly in the study of topological insulators and low-dimensional systems. Overall, this comprehensive framework bridges classical and quantum electrodynamics, providing a unified description of massive boson behavior and a foundation for future theoretical and experimental exploration in modern quantum field theory.


**Keywords:** massive electrodynamics, axion electrodynamics, relativistic quantum mechanics, 2DEG, quantum Hall effect, electromagnetic helicities, magnetoelectric effect

## 1. Introduction

A photon is a fundamental particle in nature that mediates electromagnetic interactions. Planck proposed that photons carry discrete quanta of energy, a concept that laid the foundation for quantum electrodynamics. Maxwell demonstrated that light behaves as an electromagnetic wave [1], while de Broglie extended wave-particle duality to microscopic particles, showing that photons exhibit both wave-like and particle-like properties. Electrons, on the other hand, obey quantum mechanical laws and are described by the Dirac equation that governs relativistic particles *via* an energy-momentum eigenvalue formulation. In this framework, the electron is represented by a four-component spinor wavefunction, encapsulating all possible quantum states.

The electromagnetic field of a charged particle is mathematically described by a - four-component vector potential. In quantum mechanics, physical quantities are represented by Hermitian operators to ensure real eigenvalues. The operators governing the electron's physical properties belong to the special unitary group, $SU(n)$, while those mediating electromagnetic interactions belong to the unitary group, $U(n)$. Unlike the commutative structure of $U(n)$, the non-commutative nature of $SU(n)$ plays a crucial role in defining the fundamental differences between bosons (integral spin) and fermions (half-integral spin), establishing a deep connection between group symmetries and particle properties.

A massive boson is described by the Maxwell-Proca theory [2], which modifies Maxwell's equations by introducing a mass term. The Proca equations, however, do not preserve gauge invariance. Some physicists propose that the vector potential itself acts as a wavefunction for bosons [3]. In standard quantum electrodynamics, photons, being massless, do not interact with one another, whereas the presence of mass introduces interactions among bosons.

The quaternionic formalism offers a unified approach to describing both electrons and photons by employing a generalized energy-momentum eigenvalue equation. The quaternionic wavefunction provides a broader framework for unifying interactions between spinor particles (electrons) and gauge bosons (photons), allowing for a richer description of quantum interactions.

In this chapter, we explore the electrodynamics of massive bosons using the quaternionic momentum eigenvalue equation. Our analysis reveals that a massive boson can exhibit a vanishing magnetic field while maintaining a nonzero electric field. Under different conditions, both electric and magnetic fields may persist, suggesting that the electromagnetic properties of a boson are inherently linked to its mass. This finding has practical implications, particularly in scenarios where light can induce magnetization in materials, opening avenues for new technological applications.

A fundamental question that arises is under what conditions photons exhibit mass. In superconductors, electromagnetic interactions become short range due to the Meissner effect, effectively imparting mass to bosons [4]. Inside superconductors, massive bosons behave as a superfluid, and their electromagnetic fields satisfy the Klein-Gordon equation. Additionally, the mass of the boson introduces a novel Maxwellian current, leading to a classical Hall conductivity in the absence of an external magnetic field, mirroring the quantum Hall effect. The electrodynamics of massive bosons also draws parallels with axion electrodynamics, as proposed by Frank Wilczek [5], and is closely related to the electrodynamics of topological insulators [6].

Our formulation of Maxwell's equations for massive bosons predicts that, when a static magnetic field is applied to a two-dimensional electron gas (2DEG), a quantum voltage, current, and inductance emerge. Moreover, these bosons acquire nonzero electric and magnetic fields, contributing to novel transport phenomena.

This chapter is structured as follows: In Section 2, we investigate the electromagnetic properties of massive bosons, drawing analogies to quantum particle dynamics.

Section 3 explores electrodynamics within a quaternionic mass framework, deriving self-electric and magnetic fields for massive bosons. Section 4 formulates the modified Maxwell's equations for massive bosons and compares them with Proca theory, highlighting physical implications. We also derive a formula for the capacitance per unit area of a 2DEG, as initially obtained by Luryi. In Section 5, we introduce a three-dimensional Hall-like effect and compare it to Halperin's model. Finally, we conclude with a discussion of key findings and future research directions.

## 2. Electromagnetic properties of massive bosons

The momentum eigenvalue equation for a quantum particle of mass $m_0$ in conventional quantum mechanics is given by

$$\hat{P}\psi = p\psi, \tag{1}$$

where $p$ is the momentum eigenvalue and $\psi$ is the particle wavefunction. In the quaternionic formalism, the eigenvalue equation is generalized to

$$\tilde{P}^*\,\tilde{\Psi} = p\tilde{\Psi}, \qquad \tilde{P}^* = \left(\frac{i}{c}\,\hat{E}, -\hat{\vec{p}}\right), \qquad \tilde{\Psi} = \left(\frac{i}{c}\,\psi_0, \vec{\psi}\right), \qquad p = m_0 c, \tag{2}$$

where $\hat{E}$ and $\hat{\vec{p}}$ denote the energy and momentum operators, respectively. The relativistic energy-momentum eigenvalue equation is expressed by Dirac as

$$\gamma^\mu p_\mu \psi = m_0 c \psi,$$

with $\gamma^\mu$ being the Dirac matrices and $\psi$ a spinor. Interestingly, the quaternionic analogue of the Dirac equation leads to Maxwell's equations [7].

By applying the quaternion product rule in Eq. (2), one obtains

$$\frac{1}{c^2}\frac{\partial \psi_0}{\partial t} + \vec{\nabla}\cdot\vec{\psi} + \frac{m_0}{\hbar}\,\psi_0 = 0, \tag{3}$$

$$\frac{\partial \vec{\psi}}{\partial t} + \vec{\nabla}\psi_0 + \frac{m_0 c^2}{\hbar}\,\vec{\psi} = 0, \tag{4}$$

$$\vec{\nabla}\times\vec{\psi} = 0. \tag{5}$$

Eqs. (3)–(5) have been utilized to describe the electron [8, 9]. These equations exhibit symmetry; that is, they remain invariant under the transformations

$$\vec{\psi}' = \vec{\psi} - \vec{\nabla}\lambda, \qquad {\psi_0}' = \psi_0 + \frac{\partial \lambda}{\partial t} + \frac{m_0 c^2}{\hbar}\,\lambda, \tag{6}$$

where $\lambda$ is a function satisfying eq. (10) (see below). Eq. (A) is referred to as the matter gauge transformation.

In what follows, we consider the case of a massive boson acting as a definite particle. While electromagnetic theory traditionally describes a single boson as having zero charge, zero mass, and null electric and magnetic fields, our formalism allows a single massive boson to be represented by the wavefunctions $\vec{\psi} = \vec{A}$ and $\psi_0 = \varphi$. With this identification, Eqs. (3)–(5) become

$$\frac{1}{c^2}\frac{\partial\varphi}{\partial t} + \vec{\nabla}\cdot\vec{A} + \frac{m_\gamma}{\hbar}\varphi = 0, \tag{7}$$

$$\frac{\partial\vec{A}}{\partial t} + \vec{\nabla}\varphi + \frac{m_\gamma c^2}{\hbar}\vec{A} = 0, \tag{8}$$

$$\vec{\nabla}\times\vec{A} = 0. \tag{9}$$

These equations imply that

$$\vec{E} = \frac{m_\gamma c^2}{\hbar}\vec{A}, \qquad \vec{B} = 0, \qquad \frac{1}{c^2}\frac{\partial\varphi}{\partial t} + \vec{\nabla}\cdot\vec{A} = -\frac{m_\gamma}{\hbar}\varphi. \tag{10}$$

It is noteworthy that a massive boson exhibits a vanishing magnetic field while maintaining a nonvanishing electric field, and the Lorenz gauge condition is not fulfilled. In contrast, a massless boson possesses zero electric and magnetic fields and satisfies the Lorenz gauge condition.

In the context of the photoelectric effect, where a photon interacts with an electron and behaves as a particle, one may ask: What is the force exerted by the photon on the electron at that moment? From Eq. (9), the force is given by

$$\vec{F} = \frac{q_e m_\gamma c^2}{\hbar}\vec{A},$$

with $q_e$ being the electron charge. Consequently, the electron is accelerated by

$$\vec{a}_e = \frac{m_\gamma q_e \hbar}{m_e c^2}\vec{A},$$

a value that is extremely small. For the massive boson, the gauge transformations of eq. (A) become

$$\vec{A}' = \vec{A} - \vec{\nabla}\lambda, \qquad \varphi' = \varphi + \frac{\partial\lambda}{\partial t} + \frac{m_\gamma c^2}{\hbar}\lambda. \tag{11}$$

These transformations reduce to the conventional gauge transformations (cf. Eq. (17)) when the boson is massless. A massive boson satisfying a quaternionic Dirac eigenvalue equation has been recently investigated [10].

The vector and scalar potentials for the massive boson satisfy the Telegraph equation —which governs the propagation of electrical signals in transmission lines—given by

$$\frac{1}{c^2}\frac{\partial^2\varphi}{\partial t^2} - \nabla^2\varphi + \frac{2m_\gamma}{\hbar}\frac{\partial\varphi}{\partial t} + \left(\frac{m_\gamma c}{\hbar}\right)^2\varphi = 0, \qquad \frac{1}{c^2}\frac{\partial^2\vec{A}}{\partial t^2} - \nabla^2\vec{A} + \frac{2m_\gamma}{\hbar}\frac{\partial\vec{A}}{\partial t} + \left(\frac{m_\gamma c}{\hbar}\right)^2\vec{A} = 0. \tag{12}$$

According to Eq. (9), the electric field of the massive boson also obeys Eq. (10). This field is a damped oscillatory field whose amplitude decays over time due to the boson mass; it varies as

$$\vec{E} \propto e^{-m_\gamma c^2 t/\hbar} e^{-i\left(\omega t - \vec{k}\cdot\vec{r}\right)},$$

and its spatial extent is consequently limited. Moreover, Eq. (9) ensures that $\vec{\nabla} \times \vec{E} = 0$.

The energy conservation equation for massive bosons can be derived from Eqs. (6)–(8) and is expressed as

$$\frac{\partial u}{\partial t} + \vec{\nabla} \cdot \vec{S} = -\frac{u}{\tau}, \qquad u = \frac{A^2}{2\mu_0 \lambda^2} + \varepsilon_0 \frac{\varphi^2}{2\lambda^2}, \qquad \vec{S} = \frac{\varphi \vec{A}}{\mu_0 \lambda^2}, \qquad \tau = \frac{\hbar}{2m_\gamma c^2}, \tag{13}$$

where $\vec{S}_0$ and $u$ represent the energy flux and energy density, respectively, and $\lambda = \hbar/(mc)$. The energy carried by the massive boson propagates along the direction of the magnetic vector potential, indicating that it behaves as a longitudinal wave—analogous to acoustic waves in plasmas. This characteristic may be attributed to the mechanical properties of bosons.

It is also apparent from Eqs. (11) and (9) that the energy flow is directed along the electric field. Interestingly, these equations suggest that a massless boson does not radiate. As an additional remark, Schrödinger demonstrated that an electron undergoes a rapid, jittery motion (Zitterbewegung) when moving freely [11]. The frequency of this motion is given by $\omega_z = 2m_e c^2/\hbar$; it appears that the boson oscillates at the same frequency, with $f = 1/\tau$.

The momentum conservation equation for massive bosons is

$$\frac{\partial p_i}{\partial t} = \partial_j \sigma_{ij} - \frac{p_i}{\tau}, \qquad p_i = \frac{S_i}{c^2}, \qquad \sigma_{ij} = -A_i A_j - \frac{\delta_{ij}}{2}\left(\frac{\varphi^2}{c^2} - A^2\right), \tag{14}$$

where $p_i$ is the momentum density. Recall that in the canonical quantization of the electromagnetic field, the conjugate momentum associated with the vector potential is equal to the electric field. The motion of the electromagnetic field thus bears resemblance to fluid flow, with the term $\sigma_{ij}$ representing the stress tensor of the electromagnetic field. The damping term $-\frac{p_i}{\tau}$ functions analogously to friction, with $\tau$ denoting the relaxation time of the scattered bosons. Eq. (12) is similar in form to the Drude model describing electron motion in metals. Moreover, the stress tensor $\sigma_{ij}$ parallels that of a fluid; for a relativistic perfect fluid one has

$$\sigma^f_{ij} = \left(\rho + \frac{p}{c^2}\right) u_i u_j - p\,\delta_{ij}, \tag{15}$$

where $u$ is the fluid velocity, $\rho$ the mass-energy density, and $p$ the pressure. In the original Maxwell theory, the vector potential $A_i$ is interpreted as the velocity of a fluid (i.e., the photon). Thus, Eq. (12) highlights a fluid-like aspect of the electromagnetic field when it is composed of massive bosons.

The energy conservation equation for the electromagnetic field is given by [1].

$$\frac{\partial u_e}{\partial t} + \vec{\nabla} \cdot \vec{S}_e = -\vec{J} \cdot \vec{E}, \qquad u_e = \frac{\varepsilon_0 E^2}{2} + \frac{B^2}{2\mu_0}, \qquad \vec{S}_e = \frac{1}{\mu_0} \vec{E} \times \vec{B}, \tag{16}$$

and the momentum density of the field is given by

$$\vec{p}_e = \frac{\vec{S}_e}{c^2}. \tag{17}$$

In the energy-momentum conservation equations, Eqs. (11) and (12) assume a quantum perspective of the bosons, while Eqs. (13) and (14) treat the electromagnetic field classically (as a wave). Notably, Eqs. (11) and (13) reveal that the magnetic vector potential $\vec{A}$ aligns with the Poynting vector $\vec{S}_e$.

## 3. Massive bosons with a quaternionic mass

If we now represent the boson by a quaternionic mass, that is,

$$m_0 \to \tilde{m} = \left(im_\gamma, \vec{m}\right),$$

then Eq. (2) transforms into

$$\frac{1}{c^2}\frac{\partial\varphi}{\partial t} + \vec{\nabla}\cdot\vec{A} = 0, \qquad c\vec{m}\cdot\vec{A} + m_\gamma\varphi = 0, \tag{18}$$

$$\vec{B} = \frac{\vec{m}}{\hbar}\varphi + \frac{m_\gamma c}{\hbar}\vec{A}, \qquad \vec{E} = \frac{c^2}{\hbar}\vec{m}\times\vec{A}. \tag{19}$$

A quantum particle that exhibits dual behavior—sometimes acting as a wave and other times as a particle—necessitates the definition of two masses: one corresponding to its particle aspect and the other to its wave aspect. The proper mathematical tool to describe such a particle is a quaternionic mass.

Notably, the electric and magnetic fields of the boson emerge as quantum fields that are functions of its mass. Furthermore, for massive bosons the electric and magnetic fields are orthogonal, that is,

$$\vec{E}\cdot\vec{B} = 0.$$

Eq. (16) indicates that the magnetic field vanishes when

$$\vec{A} = -\frac{\vec{m}}{m_\gamma c}\varphi.$$

In addition, one obtains

$$\vec{m}\cdot\vec{B} = \frac{\varphi}{\hbar}M^2, \qquad \vec{A}\cdot\vec{E} = 0, \qquad \vec{m}\cdot\vec{E} = 0, \qquad M^2 = m^2 - m_\gamma^2. \tag{20}$$

Under the gauge transformation

$$\vec{A}' = \vec{A} - \vec{\nabla}f, \qquad \varphi' = \varphi + \frac{\partial f}{\partial t}, \tag{21}$$

Eqs. (15) and (16) remain invariant provided that the function $f$ satisfies a wave equation propagating at the speed of light. Specifically, applying the gauge transformation to Eqs. (15) and (16) gives

$$\frac{1}{c^2}\frac{\partial^2 f}{\partial t^2} - \nabla^2 f = 0, \qquad \vec{\nabla}\cdot\left(\vec{m}f\right) + \frac{1}{c^2}\frac{\partial}{\partial t}\left(-m_\gamma c f\right) = 0, \tag{22}$$

$$\vec{\nabla}\left(m_\gamma c f\right) + \frac{\partial}{\partial t}\left(-\vec{m}f\right) = 0, \qquad \vec{\nabla}\times\left(\vec{m}f\right) = 0. \tag{23}$$

Defining new potentials by

$$\Phi = \frac{m_\gamma c^2}{\hbar} f \quad \text{and} \quad \vec{\mathcal{A}} = -\frac{c}{\hbar}\vec{m}f,$$

Eqs. (18) and (19) then lead to

$$\frac{1}{c^2}\frac{\partial^2 \Phi}{\partial t^2} - \nabla^2\Phi = 0, \qquad \vec{\nabla}\cdot\vec{\mathcal{A}} + \frac{1}{c^2}\frac{\partial \Phi}{\partial t} = 0, \tag{24}$$

$$-\vec{\mathcal{E}} \equiv \vec{\nabla}\Phi + \frac{\partial \vec{\mathcal{A}}}{\partial t} = 0, \qquad \vec{\mathcal{B}} \equiv \vec{\nabla}\times\vec{\mathcal{A}} = 0. \tag{25}$$

Eq. (21) demonstrates that these new gauge fields result in vanishing electric and magnetic fields for the boson, even though we have not assumed that the boson is massless. In effect, the mass is "gauged out" in the new fields. It follows that, with the proper choice of gauge function, the distinction between massive and massless bosons becomes blurred, as both yield null fields under these potentials. This raises an interesting question as to whether the zero electric and magnetic fields are attributable to a vanishing mass or to the boson's charge; as shown, the latter is not a determining factor.

Furthermore, with this formulation, the gauge transformation

$$\vec{A}' = \vec{A} - \frac{\vec{m}c}{\hbar}f, \qquad \varphi' = \varphi + \frac{m_\gamma c^2}{\hbar}f, \tag{26}$$

can be interpreted as a quantum gauge transformation. Although massive electrodynamics is traditionally not invariant under gauge transformations [2], the present approach demonstrates that the electrodynamics of massive bosons can indeed preserve gauge symmetry.

## 4. Massive boson electrodynamics

The electrodynamics associated with Eqs. (15) and (16) is defined by

$$\vec{\nabla}\cdot\vec{E} = -\frac{c^2}{\hbar}\vec{m}\cdot\vec{B}, \qquad \vec{\nabla}\cdot\vec{B} = -\frac{\vec{m}}{\hbar}\cdot\vec{E}, \tag{27}$$

and[1]

[1] Using Eq. (16), the term

$-\frac{m_\gamma c}{\hbar}\vec{E} + \frac{c^2}{\hbar}\vec{m}\times\vec{B} = 0$

vanishes.

$$\vec{\nabla}\times\vec{E}=-\frac{\partial\vec{B}}{\partial t}-\frac{m_\gamma c}{\hbar}\vec{E}+\frac{c^2}{\hbar}\vec{m}\times\vec{B},\qquad \vec{\nabla}\times\vec{B}=\frac{1}{c^2}\frac{\partial\vec{E}}{\partial t}+\frac{m_\gamma c}{\hbar}\vec{B}+\frac{\vec{m}}{\hbar}\times\vec{E}. \tag{28}$$

In the presence of sources, represented by the current density $\vec{J}$ and the charge density $\rho$, and upon using Eq. (16), one obtains

$$\vec{\nabla}\cdot\vec{E}=\frac{\rho}{\varepsilon_0}-\frac{c^2}{\hbar}\vec{m}\cdot\vec{B},\qquad \vec{\nabla}\cdot\vec{B}=0, \tag{29}$$

and

$$\vec{\nabla}\times\vec{E}=-\frac{\partial\vec{B}}{\partial t},\qquad \vec{\nabla}\times\vec{B}=\mu_0\vec{J}+\frac{1}{c^2}\frac{\partial\vec{E}}{\partial t}+\frac{m_\gamma c}{\hbar}\vec{B}+\frac{\vec{m}}{\hbar}\times\vec{E}. \tag{30}$$

Eqs. (23i) and (24i) reduce to the ordinary Maxwell's equations when the additional mass-dependent terms vanish. The extra charge and current densities in Eqs. (23i) and (24i) represent the novel predictions of this extended electrodynamics. In particular, Eq. (23i) indicates that the application of a magnetic field to a system induces a charge density, a phenomenon that is observed in the Hall effect. Moreover, the corresponding induced current is observed in topological insulators.

Note that Eqs. (23) and (24) are gauge invariant. By employing Eqs. (15) and (16), these equations become

$$\vec{\nabla}\cdot\vec{E}=-\left(\frac{Mc}{\hbar}\right)^2\varphi,\qquad \vec{\nabla}\cdot\vec{B}=0, \tag{31}$$

and

$$\vec{\nabla}\times\vec{E}=-\frac{\partial\vec{B}}{\partial t},\qquad \vec{\nabla}\times\vec{B}=\frac{1}{c^2}\frac{\partial\vec{E}}{\partial t}-\left(\frac{Mc}{\hbar}\right)^2\vec{A}, \tag{32}$$

where

$$M^2=m^2-m_\gamma^2. \tag{33}$$

Notably, Eqs. (25) and (26) correspond to the Maxwell–Proca equations of massive electrodynamics [2]. In contrast, Eqs. (23) and (24) describe the quantum electrodynamics for massive bosons, generalizing standard Maxwell's equations to account for quantum effects when bosons acquire mass. This framework extends ordinary electrodynamics to include a nonzero boson mass and also generalizes Frank Wilczek's axion electrodynamics for the symmetrized (dual) case, where $\vec{m}\propto\vec{\nabla}a$ and $m_\gamma\propto\partial_t a$ [5, 12]. In such a scenario, the axion field $a$ exhibits a linear space–time dependence, that is, $a=\alpha x+\beta t$, a case that is related to Weyl semimetals [13].

Eqs. (23) and (24) are invariant under time reversal provided that $\vec{m}\to-\vec{m}$. Although they violate parity, they conserve charge conjugation (CP) symmetry; parity is restored when $m_\gamma=0$. The present formulation predicts that when a magnetic field is applied, an electric charge density is induced,

$$\rho_e = -\frac{\vec{m}}{\mu_0 \hbar} \cdot \vec{B}, \tag{34}$$

an effect commonly referred to as the magnetoelectric effect. In the Hall effect, the transverse magnetic field produces charge accumulation on opposite sides of the sample, which gives rise to the Hall voltage; hence, $\rho_e$ represents the accumulated charge density due to the applied magnetic field.

Furthermore, the additional currents appearing in Eq. (24i) can be expressed as

$$\vec{J}_m = -\frac{M^2 c}{\mu_0 m_\gamma \hbar} \vec{B} - \frac{\rho_e}{m_\gamma} c\vec{m}. \tag{35}$$

The last term in Eq. (24ii) may be rewritten as $\vec{J}_e = -m_\gamma \vec{v}$ if one sets $m_\gamma \vec{v} = c\vec{m}$, representing the current due to the motion of the charge density $\rho_e$. This current can be interpreted as the inertial current of the bosons. Remarkably, both currents are nondissipative and can be regarded as supercurrents.

## 5. Massive boson implications

Interestingly, while Wilczek electrodynamics is classical, our electrodynamics is quantum. Thus, a scalar field interacting with massless bosons in Wilczek electrodynamics is equivalent to massive electrodynamics with a vanishing magnetic current. However, while axions have not been observed in experiments, our theory here does not require any exotic particles and relies entirely on massive boson dynamics. The mass of the boson induces an electric current given by

$$\vec{J}_0 = \frac{m_\gamma c}{\mu_0 \hbar} \vec{B} + \frac{\vec{m}}{\mu_0 \hbar} \times \vec{E}. \tag{36}$$

The first current is opposite to the direction of the magnetic field, and the second current is perpendicular to the electric field. The second current is similar to Hall's current, which can be observed in some electronic devices [14]. These currents can arise in a material that has no free charges, such as an insulator. The current associated with the first term in Eq. (28) is given by

$$I_q = \frac{m_\gamma c}{\mu_0 \hbar} \Phi_B, \tag{37}$$

where $\Phi_B$ is the magnetic flux.

Notably, a static flux gives rise to a quantum inductance

$$L_q = \frac{\mu_0 \hbar}{m_\gamma c} = \frac{1}{\sigma c}, \tag{38}$$

which persists even when no current is passed through a conductor. This is due to the material nature of the boson. The minus sign is similar to that of Lenz's law associated with a time-varying flux.

Integrating the first equation in Eq. (24) over a surface gives rise to an electric voltage given by

$$V_m = \frac{Q}{2\tau} Z_0, \tag{39}$$

where $Z_0 = \mu_0 c$ is the vacuum impedance and $Q$ is the total charge enclosed by the surface.

If we assume a quantized flux instead, where

$$\Phi_B = \frac{\hbar}{e},$$

then

$$I_q = \frac{m_\gamma c}{\mu_0 e} = \frac{m_\gamma c^2}{e Z_0}. \tag{40}$$

This current may arise from the particle nature of the boson. A new quantum inductance is associated with the Hall effect of the form

$$L_H = \frac{2\pi m_e}{n_s e^2}, \tag{41}$$

where $n_s$ is the surface number density in a two-dimensional electron gas (2DEG) [15]. Recall that in the Josephson effect, the critical current and inductance are related by $L = \hbar/(2eI_c)$. Accordingly, one finds a critical current associated with the 2DEG of the form

$$I_c = \frac{n_s e \hbar}{2m_e}. \tag{42}$$

Moreover, if this inductance is due to the motion (inertia) of bosons, then one could estimate the boson mass in a 2DEG from the relation

$$m_\gamma = N_\gamma m_e, \qquad N_\gamma = \frac{\mu_0 n_s e^2 \hbar}{c m_e^2}, \tag{43}$$

by equating $L_H$ to $L_q$.

An inductance arising from electron inertia is known as *kinetic inductance*, defined for a wire of length $\ell$ and cross-sectional area $A$ as

$$L_k = \frac{m_e}{ne^2} \frac{\ell}{A}, \tag{44}$$

which could dominate in a superconductor at the nanoscale.

Thus, when a static magnetic field is turned on, the 2DEG energy stored in the system is

$$U_L = \frac{\pi n_s \hbar^2}{m_e}. \tag{45}$$

This energy is equal to the Fermi energy of the 2DEG at 0 K. The voltage associated with the static magnetic flux and the magnetic energy stored is given by

$$V_q = \frac{\pi n_s \hbar^2}{m_e e}. \tag{46}$$

The current can be excited by radiation of frequency $f_q$ given by

$$I_q = e f_q, \quad \text{so that} \quad f_q = \frac{n_s \hbar}{2m_e}. \tag{47}$$

Since a typical value of $n_s$ is $\sim 10^{15}\mathrm{m}^{-2}$, one finds that $m_\gamma \sim 10^{-6} eV/c^2$. A typical frequency of 10 GHz will suffice to induce such a current. The Joule power dissipated inside the system is

$$P_q = I_q V_q = \frac{12\pi n_s^2 \hbar^3}{m_e^2}, \tag{48}$$

which for a typical 2DEG is $P_q \sim 10^{-11} W$. These interesting phenomena should be observable in a 2DEG.

The mass $\vec{m}$ describes the mass associated with the matter-wave of the particle, and the mass $m_\gamma$ represents the mass of the particle itself. This becomes apparent when we express the total relativistic energy of an object as

$$E = mv^2 + m_0 c^2 \sqrt{1 - \frac{v^2}{c^2}}, \tag{49}$$

which represents the wave-particle duality. The first term describes the matter-wave energy content, whereas the second term describes the particle energy content. As the velocity increases, the particle energy decreases, while the matter-wave energy increases. When the velocity approaches the speed of light, the object becomes a pure wave.

The wave equations associated with Eqs. (25) and (26) are given by

$$\frac{1}{c^2}\frac{\partial^2 \vec{E}}{\partial t^2} - \nabla^2 \vec{E} + \left(\frac{Mc}{\hbar}\right)^2 \vec{E} = 0, \qquad \frac{1}{c^2}\frac{\partial^2 \vec{B}}{\partial t^2} - \nabla^2 \vec{B} + \left(\frac{Mc}{\hbar}\right)^2 \vec{B} = 0, \qquad M^2 = m^2 - m_\gamma^2. \tag{50}$$

These equations suggest that the electromagnetic field obeys the Klein-Gordon equation for a massive field of mass $M$. Proca demonstrated that a classical vector boson (photon) satisfies the Klein-Gordon equation. In this formulation, the massive boson follows the Klein-Gordon equation, where the electromagnetic field is treated as a quantum field. Notably, when $|\vec{m}| = m_\gamma$, that is, when the entire entity transitions into a wave, the electromagnetic field satisfies a massless wave equation.

The energy conservation equation associated with Eqs. (23) and (24) is [16]:

$$\vec{\nabla} \cdot \vec{S}_m + \frac{\partial u_m}{\partial t} = 0, \tag{51}$$

where

$$\vec{S}_m = \mu_0^{-1}\vec{E} \times \vec{B}, \qquad u_m = \frac{1}{2}\varepsilon_0 E^2 + \frac{B^2}{2\mu_0}. \tag{52}$$

This equation indicates that the currents appearing in Eq. (24i) are nondissipative.

The momentum conservation equation corresponding to Eqs. (23) and (24) is:

$$-\frac{\partial}{\partial t}\left(\varepsilon_0\vec{E} \times \vec{B}\right)_i + \partial_j T_{ij} = 0, \qquad T_{ij} = \left(\varepsilon_0 E_i E_j + \mu_0^{-1} B_i B_j\right) - u_m \delta_{ij}, \tag{53}$$

where $T_{ij}$ represents Maxwell's stress tensor.

By defining the electric and magnetic fields as $\vec{B} = \vec{\nabla} \times \vec{A}$ and $\vec{E} = -\vec{\nabla}\varphi - \frac{\partial\vec{A}}{\partial t}$, we obtain:

$$\vec{\nabla} \cdot \left(\vec{E} \times \vec{A} + \varphi\vec{B}\right) + \frac{\partial}{\partial t}\left(\vec{A} \cdot \vec{B}\right) = -2\vec{E} \cdot \vec{B} = 0. \tag{54}$$

Interestingly, the energy density and flux remain unaffected by the boson mass. Eq. (32) can be directly compared with Eq. (11).

Adding Eqs. (30) and (33) yields:

$$\vec{\nabla} \cdot \vec{S}'_m + \frac{\partial u'_m}{\partial t} = 0, \tag{55}$$

where

$$\vec{S}'_m = \mu_0^{-1}\vec{E} \times \vec{B} + \frac{m_\gamma c^3}{\hbar}\vec{s}, \qquad u'_m = u_m + \frac{m_\gamma c^3}{\hbar}h_m, \tag{56}$$

with

$$h_m = \varepsilon_0\vec{A} \cdot \vec{B}, \qquad \vec{s} = \varepsilon_0\left(\vec{E} \times \vec{A} + \varphi\vec{B}\right). \tag{57}$$

These equations indicate that the helicity ($h_m$) and spin ($\vec{s}$) densities of the boson are linked to its mass. Consequently, the energy and momentum conservation equations differ from those of a massless boson. The additional term in the energy conservation equation is associated with the material aspects of the electromagnetic field and is inherently quantum in nature. In contrast, the classical term reflects the wave characteristics of a massless boson and vanishes when the boson becomes massless.

Since $h_m$ can assume positive or negative values, bosons can propagate in both forward and backward directions. This longitudinal energy component emerges when the boson acquires mass, contributing to the mechanical aspect of the total electromagnetic energy. While the transverse energy causes charged particles to oscillate, the longitudinal (mechanical) energy results in particle displacement. Eq. (35) reveals that matter energy propagates along the spin direction.

Eqs. (30) and (33) suggest that additional spin and helicity densities in the electromagnetic field lead to spin current conservation, defined as:

$$\vec{\nabla} \cdot \vec{s}'_m + \frac{\partial h'_m}{\partial t} = 0, \qquad \vec{s}'_m = \frac{\hbar}{m_\gamma c^3}\vec{S}_m + \vec{s}, \qquad h'_m = h_m + \frac{\hbar}{m_\gamma c^3}u_m. \tag{58}$$

Thus, the helicity conservation equation in Eq. (36) introduces new spin and helicity densities:

$$\vec{s}_A = \frac{\hbar}{m_\gamma c^3}\vec{S}_m, \qquad h_A = \frac{\hbar}{m_\gamma c^3}u_m, \tag{59}$$

when the boson possesses a nonzero mass. The additional spin component aligns with the propagation direction of the electromagnetic wave. When the canonical energy-momentum tensor of the electromagnetic field is symmetrized, an extra surface term emerges, which relates to the spin conservation equation.

Interestingly, Eqs. (42) and (43) were recently found [17] to represent Maxwell's equations of dual bosons. Eq. (43) implies that a current can be generated by applying stationary electric and magnetic fields to matter that does not even have free electrons.

The electrodynamics of topological insulators studied by Wu *et al.* can be connected with Eqs. (42) and (43) if one defines the phase angle,

$$\theta = -\frac{\pi\vec{p}\cdot\vec{r}}{\alpha\hbar} + \frac{2m_\gamma c^2}{\alpha\hbar}t, \tag{60}$$

where $\alpha$ is the fine-structure constant [6]. A comparison of the two systems of equations reveals that

$$\dot{\theta} = \frac{2c}{r_\gamma}, \quad \vec{\nabla}\theta = -\frac{\pi}{r_\gamma}\frac{\vec{m}}{m_\gamma}, \quad r_\gamma = \frac{ke^2}{m_\gamma c^2}. \tag{61}$$

An educated guess suggests that $\dot{\theta} = \omega_c$, which is the cyclotron frequency of an electron subjected to a transverse magnetic field, *that is*, $\omega_c = \frac{eB}{m_e}$. This provides a value of the boson mass depending on the magnetic field:

$$m_\gamma c^2 = eV_B, \quad V_B = \frac{r_e c}{2}B, \tag{62}$$

where $r_e$ is the classical electron radius. Equating the boson mass in Eq. (43b) with that in Eq. (28a) gives

$$B = \frac{4n_s h}{e}. \tag{63}$$

This intriguing relationship connects the surface number density to the applied magnetic field. It relates $n_s = eB/h$ to the number of states in each Landau level per unit area, denoted by the filling factor $\nu = hn_e/eB$. This indicates that Eq. (43c) amounts to $\nu = 1/4$. The quantum potential, $V_q$, using Eq. (43c), can be expressed as

$$V_q = \frac{\hbar}{4m_e}B. \tag{64}$$

This implies that $V_B = 2\alpha V_q$. In the classical Hall effect, we have $V = \frac{I}{net}B$, where $n$ and $t$ are the electron number density and sample thickness, respectively [18]. Comparing this with Eq. (43d) gives $I = \frac{n_s e\hbar}{4m_e}$, with $n_s = nt$. Thus, at the quantum level (strong magnetic field), the Hall current is given by $I_H = \frac{n_s e\hbar}{4m_e}$.

We propose a new definition of the electric and magnetic fields for massive bosons, incorporating both ordinary and new components in Eq. (16). Whether massive bosons follow this dynamic remains an open question for experimental and theoretical investigations. For a magnetic field of $1.0\,T$, a quantum voltage of $V_q = 29\mu V$ arises in the 2DEG system. A quantum capacitance associated with $V_q$, using Eq. (28b), is given by

$$C_q = \frac{Nm_e e^2}{\pi n_s \hbar^2}, \quad C_q = \frac{m_e e^2 A}{\pi \hbar^2}, \tag{65}$$

where $N$ and $A$ are the number of electrons and the system's area, respectively. The quantum capacitance per unit area is conveniently written as

$$c_q = \frac{m_e e^2}{\pi \hbar^2}, \tag{66}$$

which is a fundamental characteristic of 2DEG systems and can manifest in various physical contexts. The quantum capacitance per unit area in a 2DEG system, as derived by Luryi, is given by [19].

$$c_L = \frac{g_v m_e e^2}{\pi \hbar^2}, \tag{67}$$

where $g_v$ is a constant. While quantum capacitance can often be ignored in ordinary devices, it plays a significant role in nanoscale systems.

## 6. Three-dimensional Hall effect

For static electric and magnetic fields, Eq. (43) can be expressed in the following matrix form:

$$\begin{pmatrix} \frac{\partial B_z}{\partial y} - \frac{\partial B_y}{\partial z} \\ \frac{\partial B_x}{\partial z} - \frac{\partial B_z}{\partial x} \\ \frac{\partial B_y}{\partial x} - \frac{\partial B_x}{\partial y} \end{pmatrix} = \mu_0 \begin{pmatrix} J_x + \frac{m_\gamma c}{\mu_0 \hbar} B_x \\ J_y + \frac{m_\gamma c}{\mu_0 \hbar} B_y \\ J_z + \frac{m_\gamma c}{\mu_0 \hbar} B_z \end{pmatrix} + (c\hbar)^{-1} \begin{pmatrix} 0 & -p_z & p_y \\ p_z & 0 & -p_x \\ -p_y & p_x & 0 \end{pmatrix} \begin{pmatrix} E_x \\ E_y \\ E_z \end{pmatrix}. \tag{68}$$

In the absence of a magnetic field (anomalous Hall effect), Eq. (44) suggests an Ohm-like equation of the form $J_i = \sigma_{ij} E_j$, where $\sigma_{ij}$ is an antisymmetric tensor. Then, one finds:

$$\sigma_{xx} = 0, \qquad \sigma_{xy} = \frac{p_z}{\mu_0 c\hbar}, \qquad \sigma_{xz} = -\frac{p_y}{\mu_0 c\hbar}, \tag{69}$$

$$\sigma_{yy} = 0, \qquad \sigma_{yx} = -\frac{p_z}{\mu_0 c\hbar}, \qquad \sigma_{yz} = \frac{p_x}{\mu_0 c\hbar}, \tag{70}$$

$$\sigma_{zz} = 0, \qquad \sigma_{zx} = \frac{p_y}{\mu_0 c\hbar}, \qquad \sigma_{zy} = -\frac{p_x}{\mu_0 c\hbar}. \tag{71}$$

Interestingly, all the longitudinal conductivities are zero, but the transverse conductivities $\sigma_{ij}$ are antisymmetric and nonzero. This situation is found to occur *via* the quantum Hall effect. However, we assume here that no external magnetic field is present, but that boson motion along the transverse direction induces a magnetic field. Interestingly, a Hall effect without a magnetic field was recently proposed by Schade et al. [20]. The other alternative is to assume that this conductivity is a manifestation of the spin current, as shown in Eq. (38). One may thus express the transverse conductivities in terms of the wave vector, $\vec{p} = \hbar\vec{k}$, and the vacuum impedance $Z_0 = \mu_0 c$ as:

$$\sigma_{xy} = \frac{k_z}{Z_0}, \qquad \sigma_{yz} = \frac{k_x}{Z_0}, \qquad \sigma_{zx} = \frac{k_y}{Z_0}. \tag{72}$$

These are the transverse conductivities in three dimensions. Thus, the 3D Hall conductivity above can be expressed as:

$$\sigma_{xy} = \frac{2\pi}{\alpha\lambda_z}\frac{e^2}{2h}, \qquad \sigma_{yz} = \frac{2\pi}{\alpha\lambda_x}\frac{e^2}{2h}, \qquad \sigma_{zx} = \frac{2\pi}{\alpha\lambda_y}\frac{e^2}{2h}. \tag{73}$$

For a given system, $\lambda_x, \lambda_y$, and $\lambda_z$ can be associated with the dimensions of the system. Thus, $\lambda_z$ represents the perpendicular spacing between the x-y planes, and so forth. For a cubic structure where $\lambda_x = \lambda_y = \lambda_z = d$, one has a single transverse conductivity:

$$\sigma_{xy} = \frac{\pi}{\alpha}\frac{e^2}{h}\frac{1}{d}. \tag{74}$$

The 3D quantum Hall effect has been experimentally observed [21]. The basic theory of the 3D Hall effect was introduced by Halperin in 1987 [22]. His 3D Hall conductivities are given by:

$$\sigma_{ij} = \frac{e^2}{2\pi h}\sum_k \epsilon_{ijk} G_k, \tag{75}$$

where $G_k$ is the reciprocal lattice vector.

Now, one can express Eq. (47) as:

$$\sigma_{ij} = \frac{e^2}{2\pi h}\sum_k \epsilon_{ijk}\frac{\pi}{\alpha}k_k, \qquad \text{or} \qquad \sigma_{ij} = \sum_k \epsilon_{ijk}\frac{k_k}{Z_0}, \tag{76}$$

where $\epsilon_{ijk}$ is a completely antisymmetric tensor. From Eq. (46), one may write:

$$\sigma_{xy} = \frac{2\pi}{\mu_0 c\lambda_z}, \qquad \sigma_{yz} = \frac{2\pi}{\mu_0 c\lambda_x}, \qquad \sigma_{zx} = \frac{2\pi}{\mu_0 c\lambda_y}. \tag{77}$$

The above conductivities can be called transverse optical conductivities. For isotropic conductivity, one has $\lambda_x = \lambda_y = \lambda_z = d$ and $\sigma_{xy} = \sigma_{yz} = \sigma_{zx}$ so that

$$\sigma_{xy} = \frac{2\pi}{\mu_0 c d}, \tag{78}$$

which can be seen as a residual transverse conductivity, where $d$ is the thickness of the layer. It is a purely geometrical quantity that depends on the material thickness. It is thus a geometrical conductivity. When a transverse magnetic field, $B$, is applied on a conductor the electrons execute a circular motion. The classical transverse conductivity of a conductor is

$$\sigma_{xy} = \frac{ne}{B}. \tag{79}$$

Comparing eq.(50) with Eq.(51) yields a magnetic field of

$$B = \left(\frac{Z_0 ne}{2\pi}\right) d, \tag{80}$$

that is, $B \propto d$. Thus, the thickness determines the applied magnetic field. The magnetic field will be frozen (saturated) if it corresponds to a nonexisting layer in the material. Using Eq. (52), the magnetic flux inside the layer is $\Phi_B = \frac{Z_0 eN}{2\pi} = N \frac{\alpha}{\pi} \frac{h}{e}$, where $N$ is the number of electrons. Recall that a unit of flux quantum is defined by $\Phi_0 = h/e$. The magnetic flux of a single electron is thus $\Phi_B = \frac{\alpha}{\pi} \frac{h}{e}$.

## 7. Concluding remarks

In this chapter, we have introduced a theoretical framework for describing massive bosons, treating them as real, well-defined particles within modified quantum electrodynamics formalism. By employing the same equations that govern massive quantum particles, we unify the characteristics of bosons within both quantum mechanics and Maxwell's theory, emphasizing their dual nature. Our findings indicate that massless bosons possess vanishing self-electric and magnetic fields, whereas massive bosons exhibit a nonzero electric field. Furthermore, we establish a connection between the electric and magnetic fields and the spin properties of massive bosons, leading to distinct representations. The formalism also supports the de Broglie hypothesis by demonstrating that a particle's momentum equals its wave momentum. A significant outcome of this work is the gauge invariance of the electrodynamics of massive bosons, which introduces a novel gauge-spin transformation. Additionally, this formulation links to Wilczek's axion electrodynamics while extending its implications. The electromagnetic field of massive bosons satisfies the Klein-Gordon equation, offering a generalized approach beyond axion electrodynamics. An intriguing discovery is the emergence of a classical Hall conductivity in the absence of an external magnetic field, akin to the conventional Hall effect. This suggests possible new experimental avenues in condensed matter physics, particularly in topological insulators and 2D electron gases. Future research should focus on experimental validation and further theoretical exploration of the properties of massive bosons.

## Acknowledgements

Researchers would like to thank the Deanship of Graduate Studies and Scientific Research at Qassim University for financial support (QU-APC-2024-9/1).

## Funding statement

No funding is allocated for this study.

## Conflict of interest

We have no conflicts of interest to disclose.

## Data access statement

No new data were generated or analyzed during this study.

## Author details

Arbab I. Arbab
Department of Physics, College of Science, Qassim University, Saudi Arabia

*Address all correspondence to: arbab.ibrahim@gmail.com

IntechOpen

© 2025 The Author(s). Licensee IntechOpen. This chapter is distributed under the terms of the Creative Commons Attribution License (http://creativecommons.org/licenses/by/4.0), which permits unrestricted use, distribution, and reproduction in any medium, provided the original work is properly cited.

## References

[1] Jackson JD. Classical Electrodynamics. New York: Wiley; 1999

[2] Proca A. Sur la Theorie ondulatoire des electrons positifs et negatifs. Journal de Physique et le Radium. 1936;**7**:347

[3] Bialynicki-Birula I. On the wave function of the photon. Acta Physica Polonica A. 1994;**86**:97

[4] Meissner W, Ochsenfeld R. Naturwiss. 1933;**21**:787

[5] Wilczek F. Two applications of axion electrodynamics. Physical Review Letters. 1987;**58**:1799

[6] Wu L, Salehi M, Koirala N, Moon J, Oh S, Armitage NP. Quantized faraday and Kerr rotation and axion electrodynamics of a 3D topological insulator. Science. 2016;**354**:1124

[7] Arbab AI. Maxwellian quantum mechanics. Optik. 2017;**136**:382

[8] Arbab AI. Derivation of Dirac, Klein-Gordon, Schrodinger, diffusion and quantum heat transport equations from a universal quantum wave equation. EPL. 2010;**92**:40001

[9] Arbab AI. The Quaternionic quantum mechanics. Applied Physics Research. 2011;**3**(2):160

[10] Arbab AI. Electric and magnetic fields due to massive photons and their consequences. Progress In Electromagnetics Research M. 2014;**34**:153

[11] Schroedinger E. Sitzungber. Preuss. Akad. Wiss. Phys.-Math. Kl. 1930;**24**:418

[12] Visinelli L. Axion-electromagnetic waves. Modern Physics Letters A. 2013; **28**:1350162

[13] Zyuzin AA, Burkov AA. Topological response in Weyl semimetals and the chiral anomaly. Physical Review B. 2012; **86**:115133

[14] von Klitzing K, Dorda G, Pepper M. New method for high-accuracy determination of the fine-structure constant based on quantized Hall resistance. Physical Review Letters. 1980;**45**:494

[15] Arbab AI. The quantum Hall's effect: A quantum electrodynamic phenomenon. Chinese Physics B. 2012; **21**:127301

[16] Arbab AI. New derivation of the spin of the electromagnetic field. Optik. 2019; **184**:436

[17] Arbab AI, Alsaawi N. Maxwell's equations of dual photon. Optik. 2019; **184**:499

[18] Hall E. On a new action of the magnet on electric currents. American Journal of Mathematics. 1879;**2**:287

[19] Luryi S. Quantum capacitance devices. Applied Physics Letters. 1988; **52**:501

[20] Schade NB, Schuster DI, Nagel SR. A nonlinear, geometric Hall effect without magnetic field. PNAS. 2019;**116**:24475

[21] Tang F et al. Three-dimensional quantum Hall effect and metal–insulator transition in ZrTe5. Nature. 2019;**569**:537

[22] Halperin BI. Possible states for a three-dimensional electron gas in a strong magnetic field. Japanese Journal of Applied Physics. 1987;**26**:1913

Chapter 2

# Study of the Influence of Electromagnetic Waves on the Atomic Structure in Engineering Environments with High Radioactive Activity

*Anthony Bassesuka Sandoka Nzao*


## Abstract

Studying the influence of electromagnetic waves on the human body is often complex, as it directly depends on various factors, including the frequency, intensity, and polarization of the wave, as well as the properties of the tissues encountered. The effects of electromagnetic fields, particularly radiofrequencies, on health have been studied for several decades. Research programs have significantly increased the number of studies on the effects of electromagnetic pollution, which causes currents in the body and energy absorption in biological tissues. These effects can manifest themselves through changes in the electromagnetic and chemical properties of biological tissues and materials, especially those used in engineering, or through chemical disturbances, significant reactions in the body in radioactive areas, including the most well-known thermal effects. In addition, there are so-called non-thermal effects, which refer to biological phenomena that can occur at levels lower than those producing thermal effects. The objective of this paper is focused on the study of the influence of electromagnetic waves applied to the atomic structure of engineering and more particularly on biological tissue. The consequences related to this influence are also presented. To achieve this objective, we implemented physical laws such as Maxwell–Ampère, Maxwell–Gauss law, Maxwell–Faraday law, Ohm's law, and the model of brain electrical activity proposed by Moris–Lecar for the mathematical modeling of the studied phenomena. The 2D simulation results successfully confirmed the consistency of our programs with the existing literature.


**Keywords:** electromagnetic waves, atomic structure of the engineer, Maxwell's equations, sources of electromagnetic radiation, electromagnetic properties

## 1. Introduction

Residents of industrialized countries are exposed to electromagnetic fields on a daily basis, both in their homes and workplaces [1, 2], where their intensity is generally higher [3]. Interactions between electromagnetic waves and the environment (fauna and flora) pose potential risks, requiring protective measures [3].

In the presence of a radioactive environment, electromagnetic radiation can cause tissue damage in humans and engineered materials, with effects that vary depending on the type of radiation and the energy delivered [4–11]. Numerous epidemiological and experimental studies have established biological effects that may compromise the long-term health of living beings [12]. The potential impact of radiofrequency electromagnetic fields on human health has always been an important research topic in the field of electromagnetic field engineering applications. Furthermore, the energy received from electromagnetic waves in a spherical brain model is very high compared to that radiated, since most of it is converted into heat [8].

This article aims to examine the influence of electromagnetic fields on the particularly unitary atomic structure of materials used in engineering, with an emphasis on the associated impacts. We have taken into account the analyses of some of the authors mentioned below to guide our work [1, 9, 12–15].

Some existing work in the literature indicates that microwave thermography techniques are widely used in medicine to monitor tissue temperature and detect electromagnetic fields [3]. Tissue temperature increases with energy deposition, proportionally to the square of the electric field. Studies show that microwave electrical pulses can induce cell death. The mechanisms of destruction vary depending on the characteristics of the pulse and its repetition frequency [16, 17].

The pulse duration is close to a nanosecond, and the applied field is about MV/cm [18]. Functional dysfunction or death of irradiated cells results from variations in membrane electrical potentials. Studies also indicate that cell death results from electrical and thermal effects, with high-density induced currents flowing in a biological system irradiated by an electromagnetic wave [18]. The use of high frequencies, beyond a few tens of gigahertz, can lead, depending on the frequency, wave intensity and exposure time, to non-thermal effects that are harmful to the health of a biological system [18–20]. Electromagnetic pollution raises concerns about its potential effects on health and the environment, although research is still ongoing. Electromagnetic waves could affect the health and structure of materials, which could compromise public health and the performance of materials in electrical, electronic, and high-radioactivity environments. Thus, a mathematical and numerical approach based on Maxwell's equations was used for this article, and it is intended to be analytical, comparative, and critical [12].

Our objective is not only to study the influence of electromagnetic waves applied to the atomic structure of engineering and more specifically to biological tissue, but also to reinforce the results available in the literature. The main consequences of electromagnetic waves on biological tissues and the atomic structure of materials, including those used in the field of engineering, are also presented.

## 2. Theoretical models

### 2.1 Concepts, definitions, and phenomenology of EMC

Electromagnetic disturbance can be defined by its pollution, compatibility, and emission or susceptibility [18–20]. This phenomenon only occurs if simultaneously a "source" of parasitic signal, a vulnerable "victim", and a coupling between the two exist (see **Figure 1**) [20]. If at least one of these elements is absent, the EMC is reestablished [21].

**Figure 1.**
*Decomposition of a CEM problem [20].*

### 2.2 Pollution sources and classifications of EMC disturbances

Electromagnetic pollution, or electrosmog, comes from various sources (mobile phones, wireless equipment, cell towers, etc.) and significantly influences areas of radioactive activity. Although research is ongoing, its potential effects on human health and the environment pose risks. Some effects are man-made, and others result from the switching of power electronics or the electrical network [22, 23].

According to the existing literature, we can classify EMC disturbances according to the following aspects:

- Frequency;
- Duration;
- coupling type [22, 24, 25];
- conduction and radiation;

Couplings are classified into two categories as shown in **Figure 2** [21, 22, 24, 25].

### 2.3 Propagation modes of electromagnetic waves

It is worth mentioning that the propagation of signals occurs in two modes: differential mode and common mode, which are not included in the coupling mode, since the signals propagate asymmetrically and symmetrically (see **Figures 3** and **4**) [22, 24, 25].

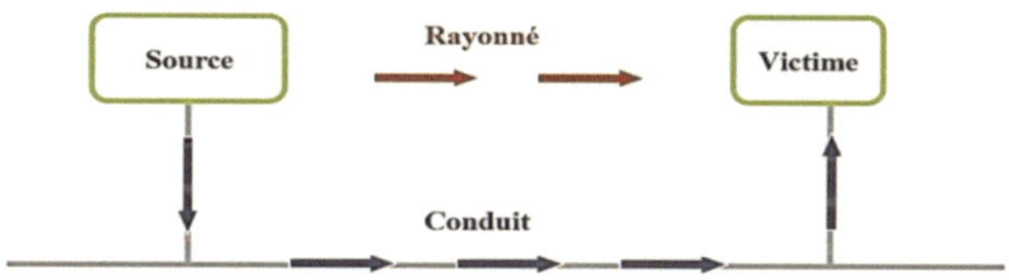


**Figure 2.**
*Coupling by radiation and conduction [20].*

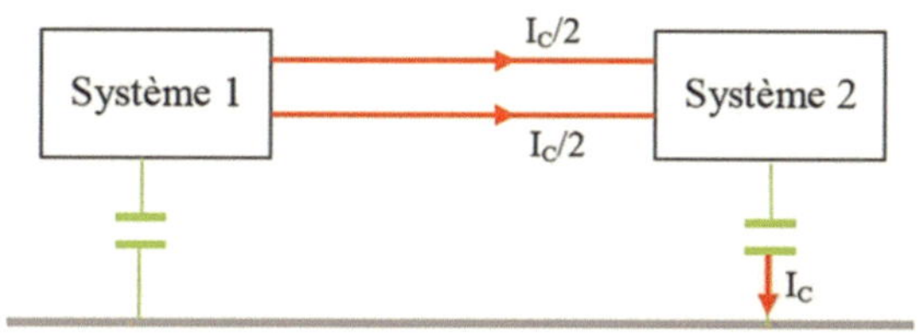


**Figure 3.**
*Common mode between two systems [20].*

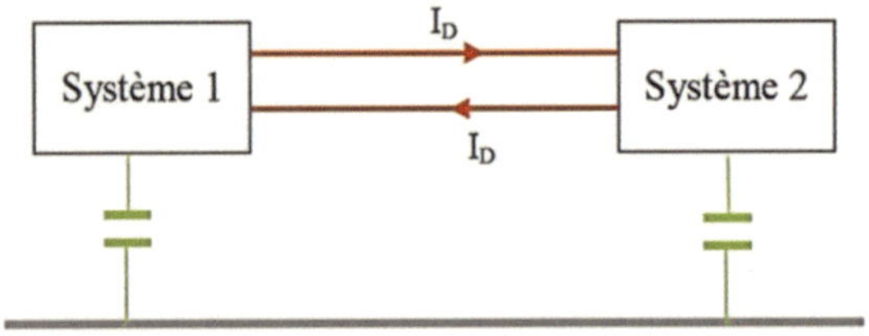


**Figure 4.**
*Differential mode between two systems [20].*

## 2.4 Influence of electromagnetic waves on the atomic structure of engineering

OEMs can have several influences on the atomic structure of materials used in engineering and several phenomena can occur that can alter the atomic structure and modify the chemical, thermal, electrical, composite and polymer design, mechanical property, durability, and reliability of the materials components including from ionization, energy level excitement, modification of chemical bonds, localized heating, impact on semiconductors, and degradation of materials. All these changes and phenomena can create enormous consequences on the environment and particularly on man. This is how a mathematical analysis has been proposed to understand the interactions of these phenomena.

### *2.4.1 Maxwell's equations*

All electromagnetic phenomena can be described by Maxwell's equations (see the Eqs. (1)–(6)). These equations allow us to link the electromagnetic field to the sources that gave rise to it. In fact, these four equations are split into two groups of two equations: The first group translates the intrinsic properties of the field (independently of the sources), and the second actually provides information on its dependence on the sources [26].

$$\text{Maxwell–Faraday} : \nabla \times \vec{E} = -\frac{\partial \vec{B}}{\partial t} \tag{1}$$

$$\text{Maxwell–Ampere} : \nabla \times \vec{H} = \vec{J}_c + \frac{\partial \vec{D}}{\partial t} \tag{2}$$

$$\text{Maxwell–Gauss} : \nabla . \vec{D} = \rho \tag{3}$$

$$\text{Ohm's law} : \vec{j} = \sigma \vec{E} \tag{4}$$

$$\vec{B} = \mu_0 \mu_r * \vec{H} \tag{5}$$

$$\vec{D} = \varepsilon_0 \varepsilon_r * \vec{E} \tag{6}$$

Relations (5) and (6) introduce the characteristics of materials to explain the effects of electromagnetic waves on the human body. Electric permittivity describes the interaction of the electric field with the dielectric material, while magnetic permeability concerns the interaction with the magnetic field. To analyze the electromagnetic problem of a radiating source and an exposed body, we need to solve Maxwell's equations [26–28].

With: E: electric field [V/m], D: field of shift electric [C/$m^2$], $\varepsilon_0$ = 8.85. $10^{-2}$ [C/Vm] the permittivity of vacuum, $\varepsilon_r$: the relative permittivity of the dielectric, ε: permittivity absolute [C/Vm], B:: field induction magnetic [Wb/$m^2$], H: field magnetic [A/m], $\mu$: permeability absolute magnetic, $\mu_0$ = 4 $\pi$.$10^{-7}$ H/m: vacuum permeability, $\mu_r$: relative permeability of a material, j: density of current (A/$m^2$), and density of charge (C/$m^3$).

### 2.4.2 Propagation of electromagnetic waves from a source

Electromagnetic waves propagate at the speed of light and in a direction opposite to their origin. This process is called radiation. A source energy [eV] of electromagnetic radiation radiates electric and magnetic fields whose spatial configuration and amplitudes differ depending on the distance between the antenna and the observation point. In general, two distinct zones are considered: the near-field zone (Fresnel zone) and the far-field zone (Fraunhofer zone). These zones are defined by the wavelength used [26–28].

### 2.4.3 Coupling path of electromagnetic waves

The atomic structure of engineering is both [26–28]:

#### 2.4.3.1 Conductive, insulating, and magnetic medium

A conductive medium is described, locally to the matter, by Ohm's law [26–28].

$$J = \boldsymbol{\sigma} \, E \tag{7}$$

With:
$\boldsymbol{\sigma}$: electrical conductivity [Ω.$m^{-1}$], J: induced current density in A/$m^2$, and E: electric field in V/m.

An insulating medium is characterized by zero conductance, therefore, no electric current [26–28]. A magnetic medium promotes the circulation of the magnetic field. It is characterized by [26–28]:

$$\mathbf{B} = \boldsymbol{\mu}\mathbf{H} \tag{8}$$

$$\boldsymbol{\mu} = \boldsymbol{\mu}_0\boldsymbol{\mu}_r \tag{9}$$

With: $\boldsymbol{\mu}_r$ Relative magnetic permeability of the medium and $\boldsymbol{\mu}_0$ Magnetic permeability of vacuum ($4\,\boldsymbol{\pi}\,\mathbf{10}^{-7}\,[\boldsymbol{H\,m}]$) .

#### 2.4.3.2 Dielectric medium

A dielectric medium promotes the circulation of the electric field. It is characterized by [26–28]:

$$\mathbf{D} = \boldsymbol{\varepsilon}\,\mathbf{E} \tag{10}$$

$$\boldsymbol{\varepsilon} = \boldsymbol{\varepsilon}_0\,\boldsymbol{\varepsilon}_r \tag{11}$$

With:
$\boldsymbol{\varepsilon}$: Dielectric permittivity also called dielectric constant, $\boldsymbol{\varepsilon}_0$ = Dielectric permittivity of vacuum ($\mathbf{10}^{-9}/\mathbf{36\pi}\,[\boldsymbol{F\,m}]$]) and $\boldsymbol{\varepsilon}_r$ Relative dielectric permittivity of the medium.

#### 2.4.4 Modeling interactions between OEMs and the atomic structure of engineering

##### 2.4.4.1 Actions and effects on the human body

Several interactions can take place when we are frequently exposed to an electromagnetic field. This leads to short-term effects, direct (biological responses), or indirect. According to the WHO, the current state of scientific knowledge does not demonstrate the long-term danger of exposure to low-intensity electromagnetic fields. The interactions between electromagnetic waves and the human body are complex and depend on a large number of factors related to the characteristics of the wave and the biological tissue encountered. An electromagnetic field comprises two components: (an electric field and a magnetic field); thus, the nature of the interactions is different for each of these components [29, 30].

##### 2.4.4.2 Interaction between electric field and living matter

###### 2.4.4.2.1 Dielectric permittivity

It describes the response of a material to an electric field. The different tissues constituting the human body each have their own permittivity. Since the body is essentially made of water, the permittivity of a large number of organs is close to that of water, which is 80 times greater than that of air. This is why the presence of a human body modifies the distribution of the electric field; this field can be defined in the air and in the human body as follows (see **Figure 5** below):

$$\vec{E} = \frac{\vec{D}}{\varepsilon_0} \text{ (In the air)} \tag{12}$$

In the presence of a human body having a permittivity $\varepsilon_r$, the field $\vec{E}$ undergoes a modification such that as demonstrated in (Eq. (13)) below:

$$\vec{E} = \frac{\vec{D}}{\varepsilon_r \varepsilon_0} \text{(In the presence of a human body)} \tag{13}$$

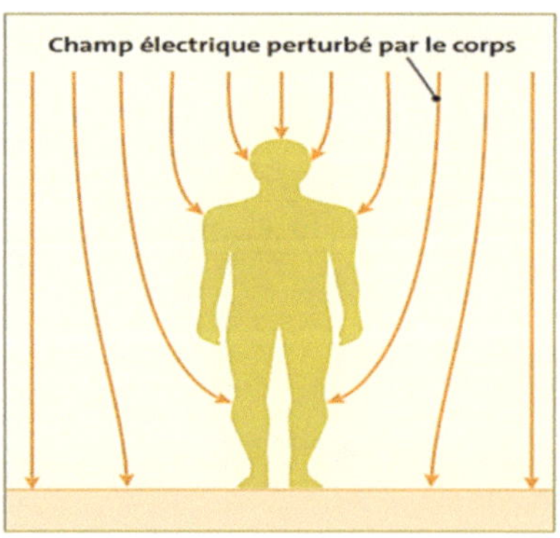


**Figure 5.**
*Perturbation of a static electric field in the presence of a body.*

#### *2.4.4.3 Creation of a current*

The electric field is the movement of electric charges within a material. According to Coulomb's law, this field can be defined as shown in (Eq. (14)) below:

$$\vec{F} = Q\vec{E}(\text{Electric force}) \tag{14}$$

(Eq. (15)) below shows that the electric field E derives from the potential V:

$$\vec{E} = -\vec{\nabla}U(\text{Field derived from potential}) \tag{15}$$

Since the conductivity of the human body is low, few charges are involved. In the case of a static field, the charges are concentrated on the body surface (see (Eq. (16)) below).

$$\vec{D} = \varepsilon_0\vec{E} + \vec{P} \tag{16}$$

With: $\vec{D}$ : *deplacement electrique en* $C/m^2$, $\vec{P}$: Electrical polarization $C/m^2$, and $\vec{E}$ : *Chamap electrique en* $V/m$.

Applying the normal vector in (Eq. (16)), we obtain (Eq. (17)) below:

$$\vec{n}\vec{D} = \varepsilon_0\vec{n}\vec{E} + \vec{n}\vec{P} \tag{17}$$

From (Eq. (17)) we derive the electric field (see (Eq. (18)) below):

$$\varepsilon_0\vec{n}\vec{E} + \vec{n}\vec{P} = \vec{n}\vec{D} \tag{18}$$

The normal of the electric field is given by (Eq. (19)) below:

$$\varepsilon_0\vec{n}\vec{E} = \vec{n}\vec{D} - \vec{n}\vec{P} \tag{19}$$

$$\vec{n}\vec{E} = \frac{\vec{n}\vec{D} - \vec{n}\vec{P}}{\varepsilon_0} \tag{20}$$

Then multiplying both sides of (Eq. (20)) by the normal vector $\vec{n}$, we obtain (Eqs. (21), (22)) below:

$$\vec{n}\vec{n}\vec{E} = \left(\frac{\vec{n}\vec{D} - \vec{n}\vec{P}}{\varepsilon_0}\right)\vec{n} \tag{21}$$

$$\text{Gold} : \vec{n}\vec{n} = 1 \tag{22}$$

So, by substituting (Eq. (22)) into (Eq. (21)), the electric field takes the form (Eq. (23)):

$$\vec{E} = \left(\frac{\vec{n}\vec{D} - \vec{n}\vec{P}}{\varepsilon_0}\right)\vec{n} \tag{23}$$

Moreover:

$$\vec{n}\vec{D} = \sigma : \text{Actual surface load;} \tag{24}$$

$$\vec{n}\vec{P} = -\sigma' : \text{Fictitious surface charge.} \tag{25}$$

By replacing expressions (24) and (25) in (Eq. (23)), we therefore obtain:

$$\vec{E} = \left(\frac{\sigma + \sigma'}{\varepsilon_0}\right)\vec{n} \tag{26}$$

Considering the relative permittivity $\varepsilon_r$ of living matter, the electric field on the surface of human skin will be:

$$\overrightarrow{E_h} = \left(\frac{\sigma + \sigma'}{\varepsilon_r \varepsilon_0}\right)\vec{n} \tag{27}$$

This equation states that in the presence of an electric field, charges accumulate on the surface of the skin. In a changing field, their orientation periodically reverses, causing an alternating movement of the charges. An alternating current is therefore created. Thus, if:

First case:

$$\vec{E} = K\frac{Q}{r^3}\vec{r} \tag{28}$$

and

$$Q = \frac{r^3}{\vec{r}}\frac{\vec{E}}{K} \tag{29}$$

The electric current generated by the movement of charges is modeled by:

$$i = \frac{dQ}{dt} \tag{30}$$

Substituting (Eq. (29)) into (Eq. (30)), we obtain:

$$i = \frac{r^2}{K}\frac{d\vec{E}}{dt} \tag{31}$$

*2.4.5 Second case*

If we apply the divergence operator to the third equation and use the partial derivative with respect to time of the second, we obtain:

$$\nabla . j = -\frac{\partial \rho}{\partial t} \tag{32}$$

This equation expresses the local conservation of charge, which results directly from Maxwell's equations. Historically, Maxwell added a term to the fourth equation to ensure this conservation. This term is similar to a current density but is not related to a mechanical motion of charges. The human body acts as a material medium for dynamic phenomena. In a dynamic system, the total current flow is conserved.

$$\vec{j} = \varepsilon_r \varepsilon_0 \frac{\partial \vec{E}}{\partial t} \tag{33}$$

(Eqs. (31) and (33)) show that the presence of a time-varying electric field induces an alternating current in the human body (see **Figure 6**).

So we can say that the higher the conductivity of a tissue, the less it opposes the passage of this current. This is why we say that the passage of such a current can be at the origin of biological effects.

#### *2.4.5.1 Penetration of electromagnetic waves into the atomic structure*

In [31], we saw that biological media are located between the dielectric and the conductor. The bases of any electromagnetic interaction with materials have been elucidated for a very long time by Maxwell's equations; however, the difficulty lies in their application to living systems. Living organisms are extremely complex and have multiple levels of organization. The propagation characteristics in biological media are as follows:

- the wavelength in a biological medium the wavelength of the signal is given by (Eq. (34)),
- the penetration depth, when a material is exposed to an electromagnetic field, it is subjected to a current density due to the movement of charges. Biological materials are not good conductors. Indeed, they conduct a current; however, the losses can be significant, and they cannot be described as lossless. Since the electromagnetic field concentrates on the surface of the conductor (weakly penetrates inside a conductor). Its penetration depth is given by Eq. (35) below;
- the energy absorbed in biological tissues, dosimetry consists of establishing the relationship between an electromagnetic field distribution in free space and the fields induced inside biological tissues or more generally the human body.

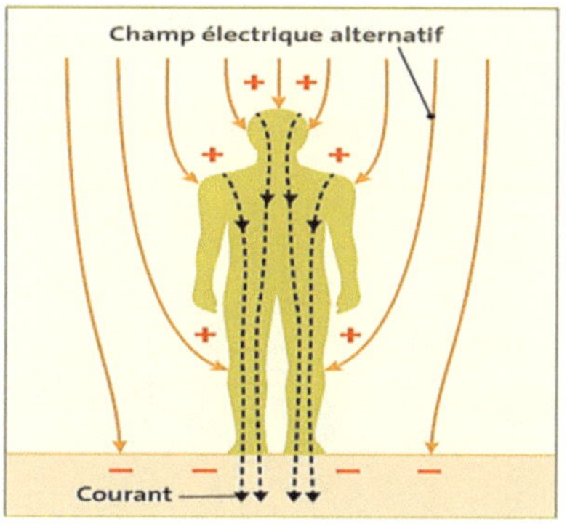


**Figure 6.**
*Creation of a current in the body due to the presence of a variable electric field.*

In other words, it is the quantification of the energy in a medium exposed to an electromagnetic field by evaluating the specific absorption rate (SAR) in the medium, which is given by Eq. (36)).

$$\lambda = \frac{\lambda_0}{\sqrt{\frac{\varepsilon^*}{2}\left[\sqrt{1+\frac{\sigma^2}{\varepsilon^2\omega^2}}+1\right]}} \tag{34}$$

$\lambda_0$ : given the wavelength in vacuum.

$$\delta = \frac{1}{\omega\sqrt{\frac{\varepsilon\mu_0}{2}\left[\sqrt{1+\frac{\sigma^2}{\varepsilon^2\omega^2}}+1\right]}} \tag{35}$$

$$\begin{cases} SAR = \frac{\sigma}{\rho}E^2 \\ SAR\,(0) = \frac{2D\,(1-R^2)}{\rho\delta} \\ SAR\,(x) = SAR\,(0).e^{-\frac{2x}{\delta}} \end{cases} \tag{36}$$

With

- $\sigma$: the conductivity of the tissue
- $\rho$: the density of the fabric
- E: amplitude of the electric field.

#### *2.4.5.2 Coupling of waves with the human body*

There are three types of coupling with the human body [32, 33]: coupling with the EBF electric field $<100\ kHz$, coupling with the EBF magnetic field, and coupling with the electromagnetic field 100 $kHz$ at 300GHz. For ELF fields, the wavelength is very long, and the time variation is very slow, so the external field induces a quasi-static field in the human body. Exposure to these waves occurs in the near-field area. Coupling with the low-frequency electric field (EBF) considering two media presented in **Figure** 7, characterized by $\sigma_1$, $\sigma_2$ and $\varepsilon_1$, $\varepsilon_2$ [34].

Noting $E_p$ and $E_\perp$, respectively, the parallel and perpendicular components of the field $E$. Following the conservation of charge on the interface, we have Eq. (37) for the EBF fields and for the static field.

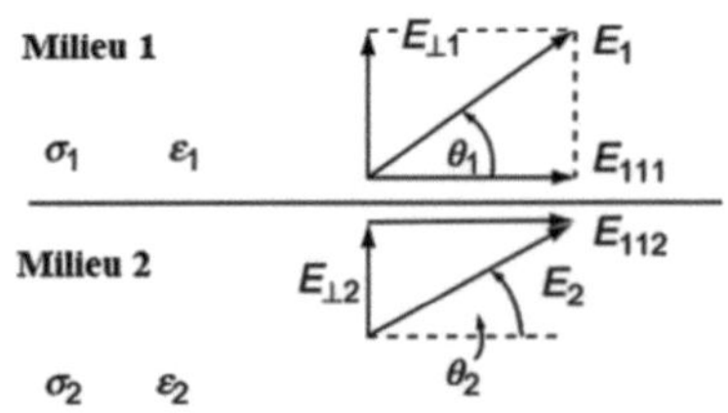


**Figure 7.**
*Passage to the E field interface.*

$$\begin{cases} E_{P1} = E_{P2} \\ \sigma_1 E_{\perp 1} = \sigma_2 E_{\perp 2} \end{cases} \tag{37}$$

The orientation of the total fields in media 1 and 2 can be described using the tangent:

$$\begin{cases} tan\,\theta_1 = \dfrac{E_{\perp 1}}{E_{P1}} \\ tan\,\theta_2 = \dfrac{E_{\perp 2}}{E_{P2}} \end{cases} \tag{38}$$

So we have:

$$tan\,\theta_1 = \frac{\sigma_2}{\sigma_1,}\,tan\,\theta_2 \tag{39}$$

For the electric field EBF, we have the (Eq. (40)):

$$\begin{cases} \sigma_1 E_{\perp 1} - \sigma_2 E_{\perp 2} = -j\omega\rho_s \\ \varepsilon_1 E_{\perp 1} - \varepsilon_2 E_{\perp 2} = \rho_s \end{cases} \tag{40}$$

$\rho_s$: being the surface current density in the ELF range, the relative permittivity of biological tissues can go up to about $10^6$, so from (Eq. (40) we have:

$$E_{\perp 1} = \frac{\sigma_2 + j\omega\varepsilon_2}{\sigma_1 + j\omega\varepsilon_1} E_{\perp 2} \tag{41}$$

This result shows that the external field must be nearly perpendicular to the biological tissue interface. Electrical charges present in an organism will be attracted to its surface or pushed as the external field alternates. Regarding the coupling with the ELF magnetic field, since the permeability of biological tissues is almost equal to that of vacuum, in the presence of the human body, the external magnetic field remains equal to the internal field (the magnetic field does not undergo a modification in the presence of the human body). Electric currents occur in a loop perpendicular to the direction of the magnetic field, since the human body is considered a conductor at low frequencies [32].

Coupling with RF (radiofrequency) and microwave electromagnetic fields, exposure to RF and microwave waves occurs in the far zone where the wave can be assimilated to a plane wave. $\eta$ being the intrinsic impedance of the medium. For a plane wave passing through a medium 1 toward a medium 2, the reflection and transmission coefficients for the case of parallel polarization and for perpendicular polarization presented in **Figure 8** are given by the relation (42) below [33, 34, 37]:

$$\begin{cases} \eta = \sqrt{\dfrac{\mu}{\varepsilon}} \\ \Gamma_{II} = \dfrac{\eta_2 \cos\theta_t - \eta_1 \cos\theta_i}{\eta_2 \cos\theta_t + \eta_1 \cos\theta_i} \\ T_{II} = \dfrac{2\eta_2 \cos\theta_i}{\eta_2 \cos\theta_t + \eta_1 \cos\theta_i} \\ \Gamma_{\perp} = \dfrac{\eta_2 \cos\theta_i - \eta_1 \cos\theta_t}{\eta_2 \cos\theta_t + \eta_1 \cos\theta_i} \\ T_{\perp} = \dfrac{2\eta_2 \cos\theta_i}{\eta_2 \cos\theta_t + \eta_1 \cos\theta_i} \end{cases} \tag{42}$$

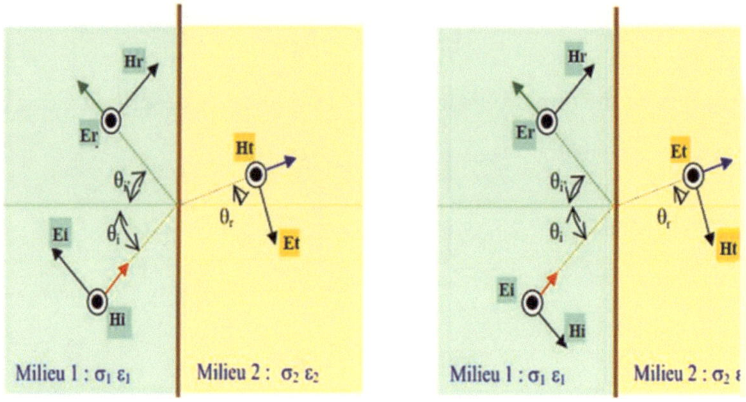


**Figure 8.**
*Polarization parallel and perpendicular to the interface [35, 36].*

### *2.4.6 Modeling the influence of OEMs on free ions of the atomic structure of engineering*

Electromagnetic waves that penetrate biological tissues will cause oscillation movements of ions. We started from the relation (43) [38] called the Lorenz force explains this phenomenon.

$$E = E_0 \sin \omega t \tag{43}$$

The force of this field exerts on each ion which crosses the plasma membrane and the transmembrane proteins. In a periodic manner (Lorentz force), it is written:

$$F_1 = E_0 z q_e \sin \omega t \tag{44}$$

This force moves the ion from its initial position. Assuming that the ion is initially at its equilibrium position, this state will be disturbed because of its displacement due to the previous force, which will receive a restoring force proportional to the displacement distance x.

$$\begin{cases} F_2 = -Dx \\ D = m_i \omega_0^2 \end{cases} \tag{45}$$

D being the restoration constant. If the moving ion experiences a moisture force that causes an attenuation of its velocity, it is defined by the following relation:

$$F_3 = -\lambda \mu \tag{46}$$

$\mu$ being the speed of the ion, the attenuation coefficient, it depends on the viscosity of the media crossed by the ion (the cytoplasm, the extracellular medium, and the protein channel) and its radius p (the ion is assumed to be spherical). Thus, the ion obtains an acceleration a, and its equation of motion is given by the following relations:

$$\begin{cases} m_i a = -\lambda \mu - Dx - E_0 z q_e \sin \omega t \\ m_i \dfrac{d^2 x}{dt^2} + \lambda \dfrac{dx}{dt} + m_i \omega_0^2 x = E_0 z q_e \sin \omega t \end{cases} \tag{47}$$

This is a second-order linear differential equation with a nonzero constant coefficient. It represents the model of the influence of OEMs on free ions of the atomic structure of engineering whose particular solution is of the form:

$$x = \frac{E_0 z q_e \cos \omega t}{\lambda \omega} \tag{48}$$

(Eq. (47)) can also be solved using some computational tools.

### *2.4.7 Modeling of thermal effects of OEMs on biological tissues or other materials*

Referring to work published in 1948 [38] by Pennes who demonstrated based on the proposed equation that includes all hypotheses of heat transfer from biological tissues to other materials including the blood perfusion rate. We considered this model because it is simple and precise, and it is written:

$$\rho C_p \frac{\partial T(t)}{\partial t} = \nabla.(k\nabla T) + \rho.SAR - B.(T - T_b) \tag{49}$$

Where: T(t) being the temperature of the tissue at time t, $\rho$: density of the tissue [kg/m$^3$], k: specific heat of the tissue [J/kg °C], k: thermal conductivity of the tissue [J/s.m.°C], and B: blood perfusion.

If we consider that the temperature variation in (Eq. (49)) is zero, we can deduce the time magnitude of the temperature as a function of the SAR (see (Eq. (50)) below):

$$\frac{\partial T(t)}{\partial t} = \frac{SAR}{C_p} \tag{50}$$

Replacing SAR with its expression in (50), we therefore obtain:

$$\frac{\partial T(t)}{\partial t} = \frac{\sigma}{\rho.C_p} E_0^2 \tag{51}$$

(Eq. (51)) represents the temperature in biological tissues or materials irradiated by the electromagnetic field.

### *2.4.8 Modeling non-thermal effects of OEMs on biological tissue*

The mechanism of wave interaction that causes these effects is based on the hypothesis that an external oscillating field exerts an oscillating force on free ions inside and outside the cell plasma membrane that can cross through it via transmembrane proteins. This force causes a forced vibration of the free ions as we have demonstrated above [39–42]. When this force reaches a threshold value, the moving ions can give false information in the control signals for opening and closing (mechanical or electrical) of the channels. This leads to the electrochemical imbalance of the plasma in the cell that will subsequently cause a disruption of the function of the entire cell. In the context of this study, we consider the electrical activity of the biological neuron [43–45].

In order to reproduce the electrical activity of the biological brain neuron, several models exist in the literature, and this article uses the mathematical formalisms based on the model of Catherine Morris and Harold Lecar [39–42] (see (Eq. (52)) below).

$$\begin{cases} \frac{d\,v(t)}{dt} = \sum_{i=1}^{I} g_i(t)(V_i - v(t)), \\ \frac{dg_i(t)}{dt} = \frac{G_i(v(t) - g_i(t))}{\tau_i(v(t), g(t)}, g_i \geq 0, i = 1, 2, \dots, I \end{cases} \tag{52}$$

With $g = (g_1, \dots, g_I)$.

The quantities $V_i$, called Nernst reversion potential, are given constants. We assume that the functions $G_i\ (t) > 0$ (the equilibrium opening rates) and $\tau_i$ (the characteristic times) are of class $C^1$, and there exist constants $0 < \tau_- < \tau_+ < \infty$ such that:

$$\tau_- \leq \tau_i(g, v) \leq \tau_+ \tag{53}$$

To describe this potential difference across the membrane [39–42]:

$$\begin{cases} I(t) = \frac{dQ(t)}{dt} \\ C\frac{dv(t)}{dt} = I_{cap}(t) \end{cases} \tag{54}$$

The constant $C = 1\ \mu F/cm^2$ will often be taken equal to 1 in the rest of this article. Then we consider that currents through the membrane result (in parallel) from the capacitance effect and the opening of ion channels (sodium $Na$, potassium $K$, and chloride $Cl^-$ associated with the leakage current). Each ionic current $I_i$ results from a resistance modulated by the opening of the channel that we describe by Ohm's law, $v = R.I$, hence the relationship, considering that a current $I_{stim}$ is applied [39–42].

$$\begin{cases} I_{cap}(t) + \sum_i I_i(t) = I_{stim} \\ I_i(t) = g_i(t)(v(t) - V_i) \\ g_i = \frac{1}{R_i} \end{cases} \tag{55}$$

We can then assume that $g_i\ (t) \equiv G_i\ (v(t))$ and reduce the complexity of the model (52). The model proposed by Catherine Morris and Harold Lecar under this hypothesis can be written in this form:

$$\begin{cases} \frac{d\,v(t)}{dt} = I_{stim} + G_L(V_L - v(t)) + G_{N_a}(v(t))(V_{N_a} - v(t)) + g_K(t)(V_K - v(t)), \\ \frac{dg_k(t)}{dt} = G_K\left(v(t) - g_K(t)\right). \end{cases} \tag{56}$$

The numerical values of the parameters $V_i$ allow us to determine the $G_i$ through the following relation:

$$G_i(v) = \frac{\tilde{G}_i}{1 + \exp\left(\frac{v_{\frac{1}{2}} - v}{k}\right)} \tag{57}$$

For this purpose, the model proposed by Catherine Morris and Harold Lecar under this hypothesis can be written in the following form:

$$\begin{cases} \dfrac{d\,[v(t)]}{dt} = (I_{stim} + I_{induit}) + G_L(V_L - v(t)) + G_{Na}(v(t))(V_{Na} - v(t)) + g_K(t)(V_K - v(t) \\ \dfrac{dg_K(t)}{dt} = G_K(v(t)) - g_K(t) \end{cases} \tag{58}$$

## 3. Simulation

### 3.1 Declaration of proposed model parameters

- For typical ions like $Na^+$ we have $\lambda = 10^{-12}\frac{\text{Kg}}{\text{sec}}$ and $m_i = 3{,}8.10^{-26}\text{Kg}$
- Electron charge: $q_e = 1{,}602.10^{-19}\text{C}$
- Electromagnetic field propagation frequency: $f$ = 50 to 100 GHz
- Electric field: $E_0 = 4$ at 10 V/m
- Density of biological tissue: ρ = 870 kg / m$^3$
- Average specific heat capacity of the fabric $C_p = 4000$ J.Kg$^{-1}$.K$^{-1}$
- Typical conductivity of biological tissue: $\sigma = 10^{-1}S/m$
- $\mu_0 = 4\,\pi\,10^{-7}$ [M/m]
- $\varepsilon_0 = 10^{-9}/36\pi$ $[F/m]$
- The parameters of the model proposed by Catherine Morris and Harold Lecar are taken from the work of Benoit Perthame [39].

### 3.2. Results and discussions

According to various 2D simulations, exposure to OEMs can have adverse effects on the atomic structure of biological tissues and materials used in engineering.

**Figure 9** shows the numerical model of the wavelength versus frequency of OEMs. **Figure 10** shows that (wavelength) is inversely proportional to f (frequency of OEMs), while **Figure 11** illustrates the block facilitating their penetration into biological materials and tissues. A high frequency limits the penetration of OEMs into biological materials and tissues, as illustrated in **Figure 12a**. Beyond 10 GHz, the depth decreases rapidly with frequency (see the work of Rakotomananjara and Randriamitantsoa [31] in **Figure 12b**)). A decrease in SAR is observed in the skin and low absorption of EM energy at 2 cm (see **Figure 13a**). This result is similar to the experiments of Rakotomananjara and Randriamitantsoa (**Figure 13b**).

In the case of heating, there is a forced vibration of ions in the atomic structure of biological tissues (see **Figure 14**). An oscillating field exerts a force on the ions of the plasma membrane (see **Figure 15**).

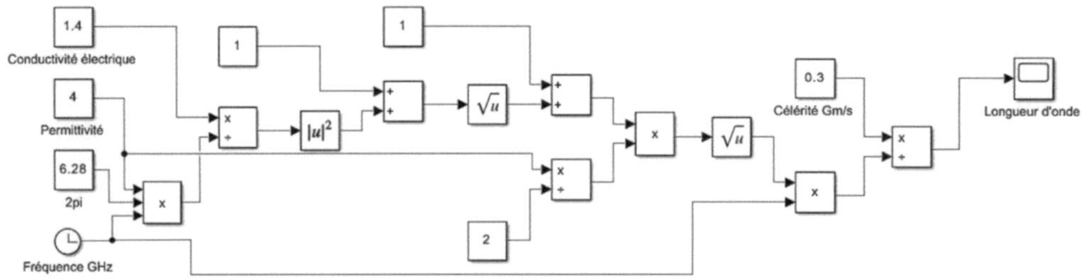


**Figure 9.**
*Numerical model of wavelength versus frequency of OEMs.*

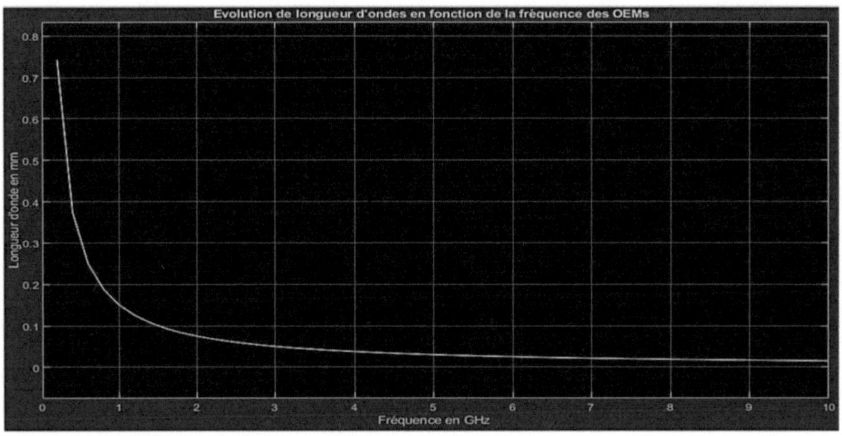


**Figure 10.**
*Behavior of wavelength as a function of OEMs frequency: Simulation result.*

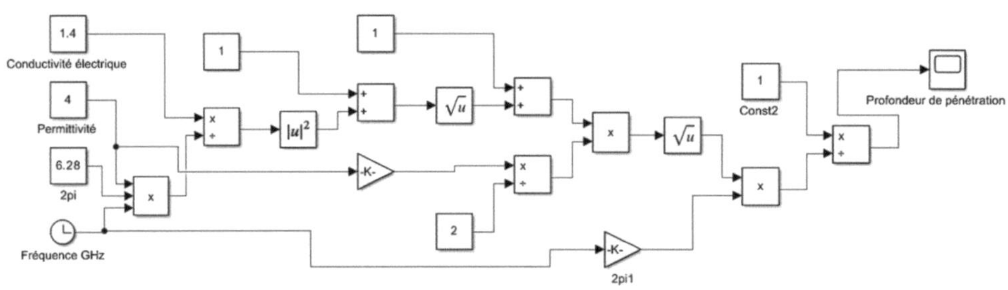


**Figure 11.**
*Numerical model of penetration depth of OEMs into biological tissue. (a) Simulation result. (b) Result published by Rakotomananjara and Randriamitantsoa [31].*

**Figure 16** presents the numerical model of the temperature rise of biological tissues under the influence of OEMs, which led to the conclusion of **Figure 17a** which explains that the temperature of biological tissues and materials increases with the time of exposure to OEMs (see also the work of Carlos Konlack and Roger TCHUIDJAN [46] presented in **Figure 17b**).

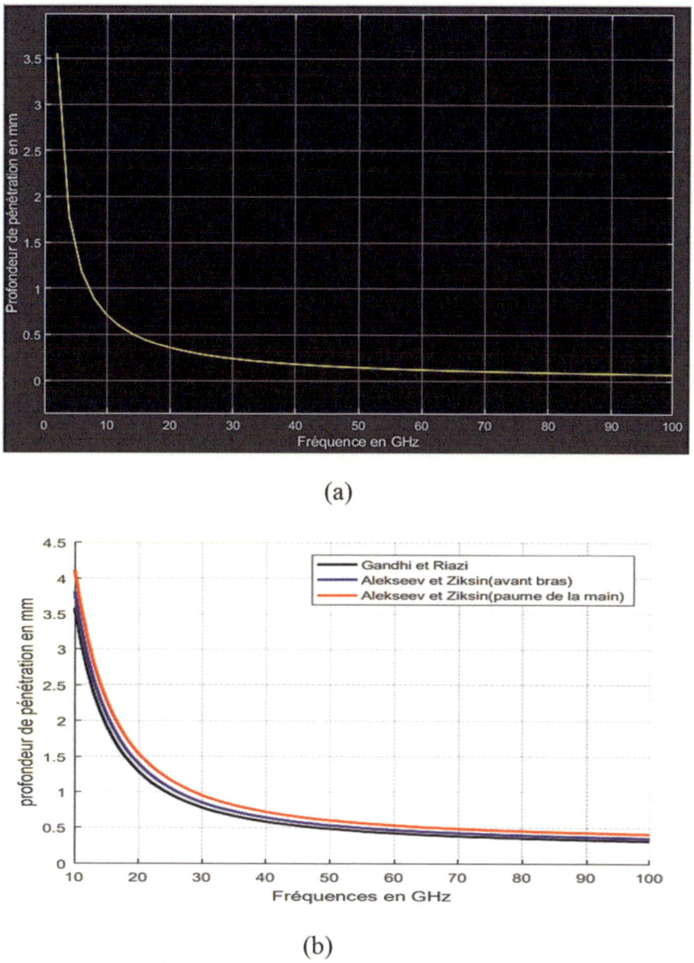


**Figure 12.**
*Penetration depth 100 MHz to 100 GHz. (a) Simulation result. (b) Result published by: Rakotomananjara and Randriamitantsoa [31].*

The numerical resolution of the electrical activity of biological neurons according to the Morris–Lecar model is illustrated in **Figure 18**.

**Figure 19a** explains the brain electrical activity without the influence of OEMs, this result shows that the excitation period of neurons is 9 seconds with a frequency of 0.11 Hz. This result is similar to the experiments of Benoit Perthame [39], see **Figure 19(b)**.

**Figure 20** explains the numerical resolution block of the brain electrical activity model proposed by Morris–Lecar, including the influence of OEMs.

Finally, **Figure 21** shows that if a 190 mA current is induced by an electromagnetic field in the brain, at 0.1 seconds, the neuron excitation frequency is reduced to 10 Hz, which disrupts the brain's electrical activity cycle, which could consequently lead to adverse health consequences.

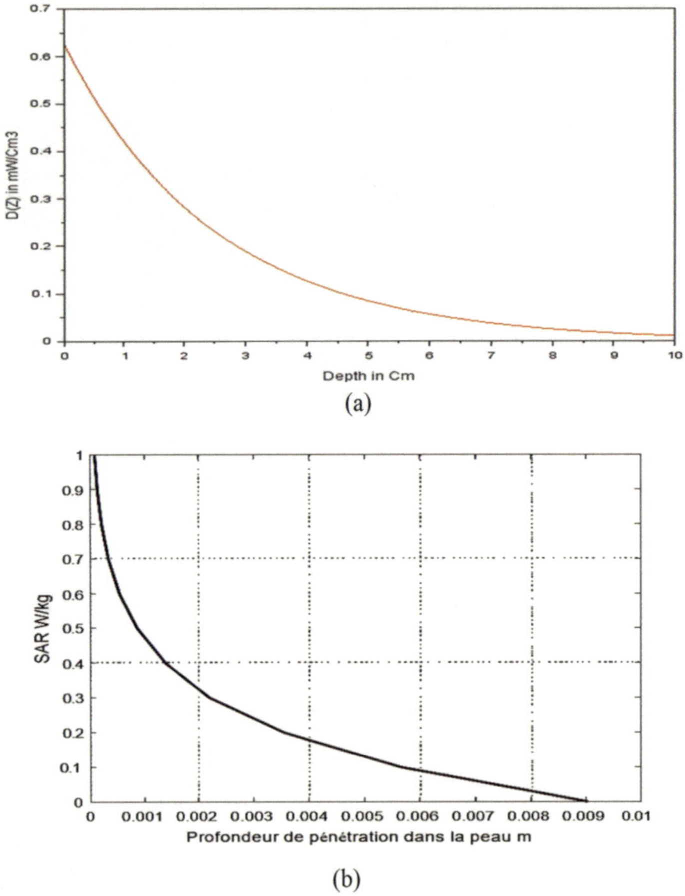


**Figure 13.**
*SAR attenuation in skin.*

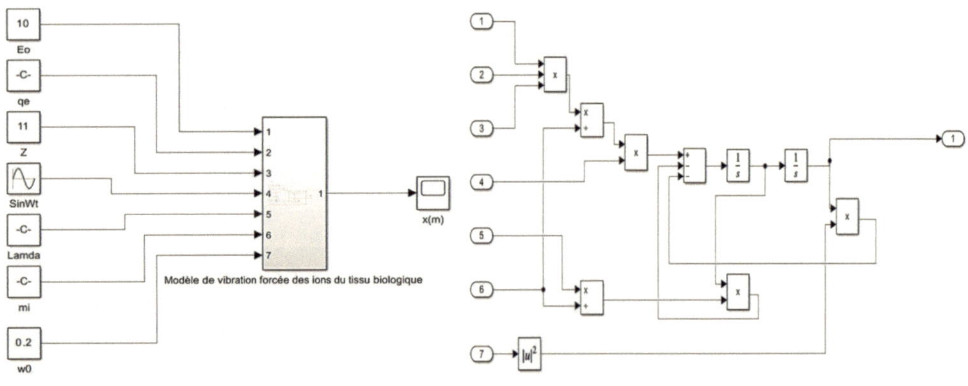


**Figure 14.**
*Numerical model of forced vibration of ions in the atomic structure of biological tissue.*

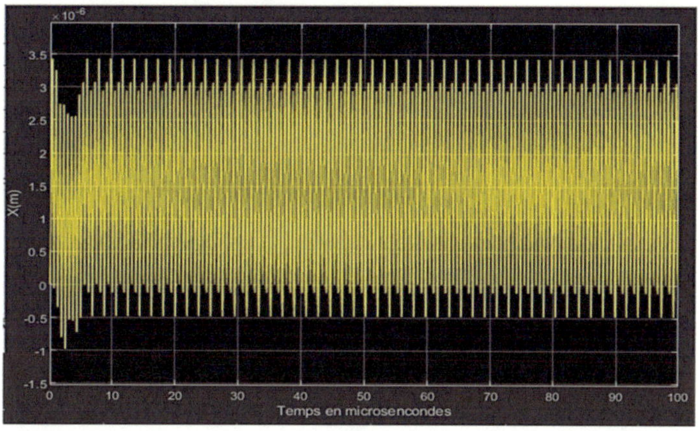


**Figure 15.**
*Forced vibration of free ions in the atomic structure of biological tissue.*

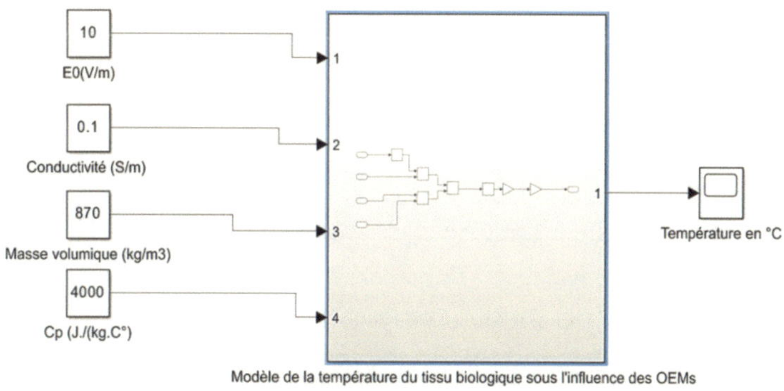


**Figure 16.**
*Numerical model of the temperature rise profile of biological under the influence of OEMs.*

## 4. Consequences of OEMs on the atomic structure of materials including those used in the engineering field

The influence of electromagnetic waves on health and the atomic structure of materials, particularly those used in engineering, can have several significant consequences:

- Heating by dielectric losses
- Structural modifications
- Ionization and effects on electrons
- Chemical bond disruption
- Electronic interference
- Applications in nanotechnology
- Resilience and security.

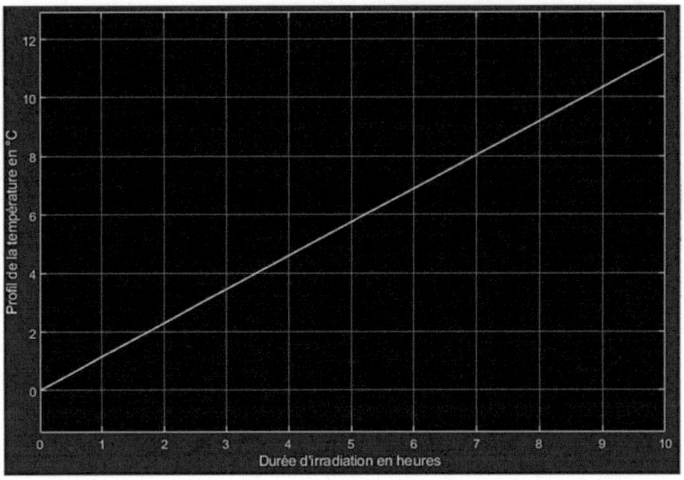


(a)

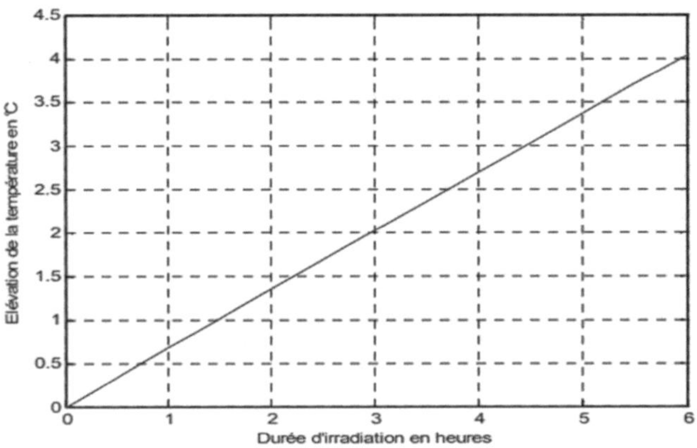


(b)

**Figure 17.**
*Temperature profile in biological tissue as a function of exposure time. (a) Simulation result (b) Result published by Carlos Konlack et Roger Tchuidjan [46].*

## 5. Conclusions

Modeling based on Maxwell's equations and the physical phenomena of EMF propagation, along with experimental values, enabled simulations of the influence of EMF on the atomic structure of the engineering in the presence of high radioactive activity.

The simulation results from the proposed models demonstrated the potential risks of EMF exposure in environments with high radioactive activity, which can lead to serious illnesses and health consequences. Above all, this is a concern when there is ecosystem modification due to pollination and interactions between species because the degradation of manufactured materials increases the risk of leaks and environmental accidents. These elements were addressed in this work not only to provide arguments to justify the relevance of the already known results but also to demonstrate and recall the importance of using various direct applications in establishing

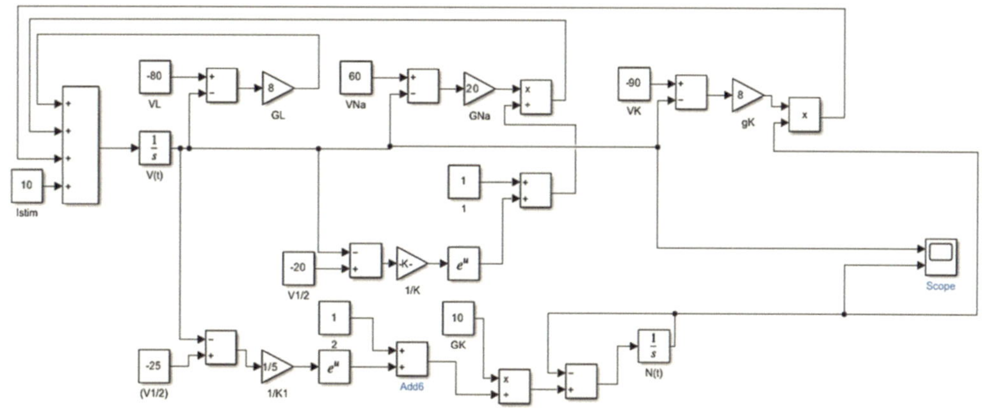


**Figure 18.**
*Simulation block of the electrical rhythm of the biological neuron.*

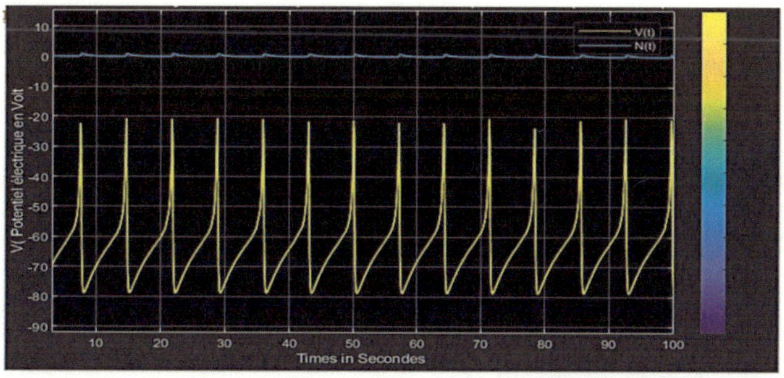


(a)

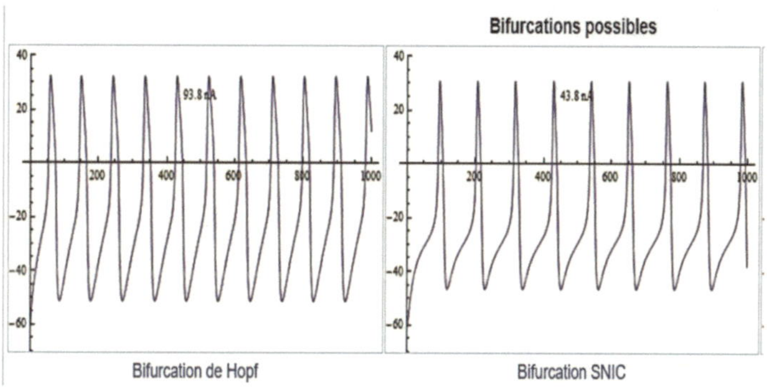


(b)

**Figure 19.**
*Electrical activity of the biological neuron without influence of the electromagnetic field (a) Simulation result (b) Result published by (Benoıt Perthame) [39].*

standards for exposure to electromagnetic fields, which are a permanent hazard, especially when crossing an area with high radioactivity density.

A study and mathematical models could still be included in our future publications because further research is needed to alert OEMs to the dangers.

Thus, comprehensive modeling will allow us to study this phenomenon based on the various mathematical models we mentioned above.

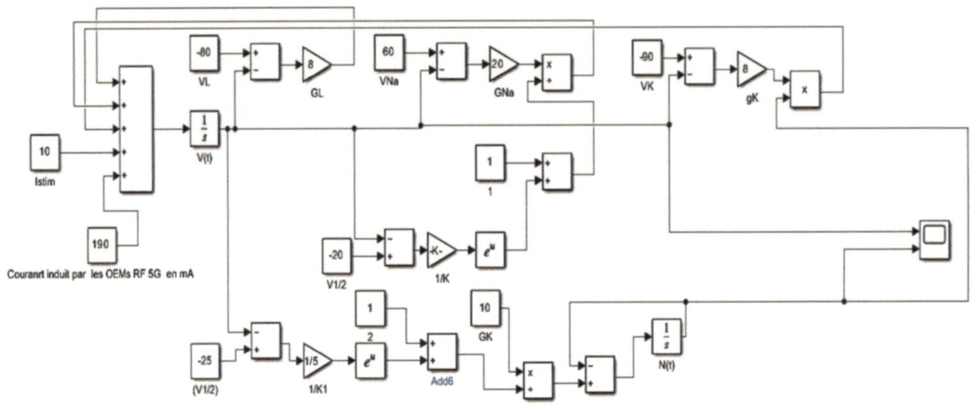


**Figure 20.**
*Simulation block of the electrical rhythm of the biological neuron with influence of the electromagnetic field.*

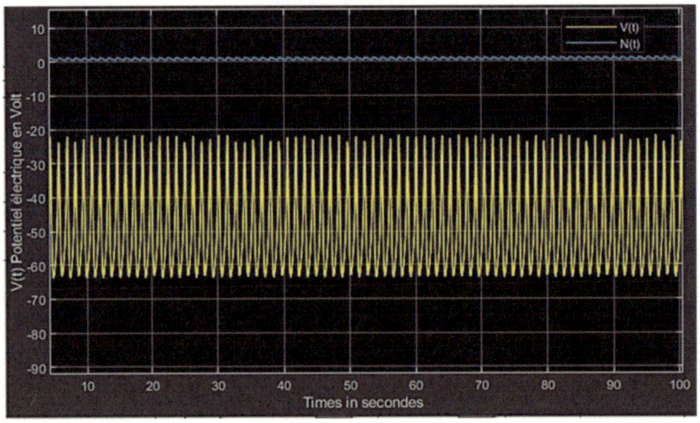


**Figure 21.**
*Simulation result of the electrical rhythm of the biological neuron under the influence of the electromagnetic field.*

## Author details

Anthony Bassesuka Sandoka Nzao
ISTA Kinshasa, Electrical Engineering, Kinshasa, Congo

*Address all correspondence to: bass_sandoka@yahoo.fr

IntechOpen

© 2025 The Author(s). Licensee IntechOpen. This chapter is distributed under the terms of the Creative Commons Attribution License (http://creativecommons.org/licenses/by/4.0), which permits unrestricted use, distribution, and reproduction in any medium, provided the original work is properly cited.

## References

[1] Advisory Group on Non-Ionising Radiation. Health Effects from Radiofrequency Electromagnetic Fields. United Kingdom: Health Protection Agency; 2012

[2] NRPB. Health effects of radiofrequency electromagnetic fields. In: Non-Ionizing Radiation (Document of the NRPB,Vol. 14, No. 2, 1-9). Didcot-Oxfordshire;

[3] French Agency for Environmental and Occupational Health Safety. Radiofrequencies: Update of the Expertise Relating to Radiofrequencies. Collective Expert Report (Referral no. 2007/007). Paris: French Agency for Environmental and Occupational Health Safety; 2009

[4] French National Agency for Food, Environmental and Occupational Health Safety. Radiofrequencies and Health: Update of the Expert Report—Anses Opinion. Collective Expert Report (Referral no. 2011-SA-0150). Maisons-Alfort: French National Agency for Food, Environmental and Occupational Health Safety; 2013

[5] Opinion of the High Council for Hygiene. Use of Wireless and GSM Communication Systems in Hospitals. In: Cuypers D, editor. Brussels: Federal Public Service for Public Health, Food Chain Safety and Environment; 2007. Available from: https://www.health.belgium.be/sites/default/files/uploads/fields/fpshealth_theme_file/19090388/Brochure_elektromagnetische_FR_kl_resolutie.pdf

[6] Vlaanderen M. Electromagnetic Fields and Health. In: Cuypers D, editor. Federal Public Service for Public Health, Food Chain Safety and Environment; 2008. Available from: https://www.health.belgium.be/sites/default/files/uploads/fields/fpshealth_theme_file/19090388/Brochure_elektromagnetische_FR_kl_resolutie.pdf

[7] Anses. Health Effects Related to Exposure to Low-Frequency Electromagnetic Fields. Alfort Cedex: French National Agency for Food, Environmental, and Occupation al Health and Safety; 2019. ISBN 979-10-286-0429-5. Available from: https://www.anses.fr/fr/system/files/AP2013SA0038Ra.pdf

[8] Anses. Exposure to Electromagnetic Fields Linked to the Deployment of "5G" Technology. Alfort Cedex: French National Agency for Food, Environmental, and Occupation al Health and Safety; 2022. Available from: https://www.anses.fr/fr/system/files/AP2019SA0006RA-2.pdf

[9] Gauthier M, Gauvin D, Environmental Health and Toxicology Department. Evaluation of the Health Effects of Electromagnetic Fields in the Field of Radio Frequencies. National Institute of Public Health of Quebec; 2016. Available from: https://mobile.inspq.qc.ca/pdf/publications/2119_evaluation_champs_electromagnetiques_radiofrequences.pdf

[10] Cousson Y. Limit Occupational Exposure to Electromagnetic Fields. INRS ed. Occupational Health and Safety – HST; 2022. Available from: https://www.inrs.fr/risks/champs-electromagnetics/ce-qu-il-faut-retenir.html

[11] Rakotomananjara DF, Randriamitantsoa PA. Modeling of the Mechanisms of the Interactions of Electromagnetic Radiation with Living Beings. Vol. 2. MADA-ETI; 2020. ISSN 2220-0673. Available from: http://

madarevues.recherches.gov.mg/IMG/pdf/art_no11_2020_vol_2_pp_110-127_modelisation_des_mecanismes_des_interactions_des_rayonnements_electromagnetiques_avec_les_et.pdf

[12] SRPE: European Parliamentary Research Service. The Effects of 5G Wireless Communication on Human Health. Policy Department for Economic, Scientific and Quality of Life Policies, Directorate-General for Internal Policies, European Parliament; 2020. Available from: https://www.europarl.europa.eu/RegData/etudes/BRIE/2020/646172/EPRS_BRI(2020)646172_EN.pdf

[13] Dos Santos M. Modeling of the topology of energy deposits created by ionizing radiation at the nanoscale in cell nuclei and its relationship to radiation-induced early events [doctoral thesis]. University of Bordeaux I Doctoral School of Physical and Engineering Sciences Specialization: Astrophysics, Plasmas, and Nuclear Energy; 2013. Available from: https://tel.archives-ouvertes.fr/tel-00931869/document

[14] Nzao ABS. Study and Modeling of human biological tissues exposed to high-frequency electromagnetic. Open Journal of Applied Sciences. 2021;**11**(10). DOI: 10.4236/ojapps.2021.1110083

[15] Nzao ABS. FDTD analysis and Modeling of microwave electromagnetic wave influences on human biological systems. Open Journal of Applied Sciences. 2022;**12**(6). DOI: 10.4236/ojapps.2022.126063

[16] Lanovici M, Morf J-J. Electromagnetic Compatibility. Switzerland: Polytechnic and University Presses Romandes; 1979

[17] Cocquerelle J-L. EMC and Power Electronics. Paris: Editions Technip; 1999

[18] Degauque P. Electromagnetic Compatibility: Noise and Radioelectric Disturbances. Dunod ed. Paris: Technical and Scientific Collection of Telecommunications; 1990

[19] Zangui S. Determination and modeling of the magnetic near-field coupling between complex systems [doctoral thesis]. Ecole Centrale de Lyon; 2011. Ed. HAL Open Science; 2012. Id: tel-00681821. Available from: https://theses.hal.science/tel-00681821v1

[20] Ikhlef N. Electromagnetic radiation of the energy transport network: Coupling with wire structures, reduction of the magnetic field [Master's thesis]. Algeria: Electrotechnics from the Unversity of Jijel; 2002

[21] Weens Y. Modeling of energy cables subjected to constraints generated by power electronics converters [thesis]. University of Science and Technology of Lille; 2006. Ed. Lab. LE2P of Lille 1. Available from: https://l2ep.univ-lille.fr/fileupload/file/theses/TheseWEENS.pdf

[22] Beltramini M. Contribution to the optimization of the converter/output filter assembly with respect to aircraft EMC constraints [thesis]. University of Toulouse; 2011. Ed. profile-feym-2012. Available from: https://theses.fr/2003DENSA020

[23] Frantz G. System approach for the study of electromagnetic compatibility of on-board networks [thesis]. University of Grenoble; 2015. Ed. HAL Science. Id: tel-01218578. Available from: https://theses.hal.science/tel-01218578v1/file/FRANTZ_2015_archivage.pdf

[24] Vermaelen C. Contribution to the modeling and reduction of conducted disturbances in variable speed drive

systems Doctoral thesis defended the Cachan Higher Normal School, 2003. Ed. thèse.fr. Available from: https://theses.fr/2003DENSA020

[25] Melit M. Study of disturbances in power electronics converters [thesis] Blaise Pascal University – Clermont II, 2008. Ed. HAL Science; 2012. Id: tel-00728780. Available from: https://theses.hal.science/tel-00728780/document.

[26] Aines I, Asma B. Analytical modeling of electromagnetic fields for the estimation of low-frequency public and professional exposure [Master's thesis field]. Annaba: Science and Technology SECTOR: Electrical Engineering Specialty: Electrical networks, University of Badji Mokhtar; 2019

[27] Delaballe J. Technical Notebooks, EMC: Electromagnetic Compatibility. Technical Notebook No. 149 Schneider Electric Edition. 2001. Available from: https://eduscol.education.fr/sti/sites/eduscol.education.fr.sti/files/ressources/techniques/3398/3398-ct149.pdf

[28] Selsabil BD. Calculation of protections for an HTA departure (30 KV) [thesis] (Academic Master's Degree). Algeria: Kasdi Marbah Ouargla University; 2014. Ed.thesis-Algeria. Available from: https://theses-algerie.com/1440645407249192/memoire-de-master/universite-kasdi-merbah—ouergla/calcul-des-protections-d-un-d%C3%A9part-hta-30-kv-

[29] Scorretti R. Numerical and experimental characterization of the LF magnetic field generated by electrotechnical systems with a view to modeling the currents induced in the human body [PhD thesis]. Lyon: Lyon Central School; 2003

[30] Purschke M, Laubach H-J, Rox Anderson R, Manstein D. Thermal injury causes DNA damage and lethality in surrounding unheated cells: Heat spectator effect active. Journal of Investigative Dermatology. 2010;**130**: 86-92. DOI: 10.1038/jid.2009.205

[31] Rakotomananjara DF, Randriamitantsoa PA. Modeling of the mechanisms of interactions of electromagnetic radiation with beings, telecommunications, automatic, signal and images research laboratory (LR-TASI). In: Doctoral School in Sciences and Techniques of Engineering and Innovation. Vol. 2. University of Antananarivo; 2020. Ed. MADA-ETI, ISSN 2220-0673

[32] Moureaux P. Exposure to Electromagnetic Fields. Ed. INRS; 2020. Available from: file:///C:/Users/user1/Downloads/guide-oseray.pdf

[33] Conil E. Electromagnetics in Complex Medium: From the Near Field to the Far Field. National Polytechnic Institute of Grenoble; 2005. Ed. HAL Id: tel-00011992; 2006. Available from: https://theses.hal.science/tel-00011992v1

[34] Neubauer G, Feychting M, Hamnerius Y, Kheifets L, Kuster N, Ruiz I, et al. Feasibility of future epidemiological studies on possible health effects of mobile phone base stations. Bio Electromagnetics. 2007;**28** (3):224-230

[35] Pozar DM. Microwave Engineering. 4th ed. Amherst: University of Massachusetts; 2012. Available from: http://mwl.diet.uniroma1.it/people/pisa/RFELSYS/MATERIALE%20INTEGRATIVO/BOOKS/Pozar_Microwave%20Engineering(2012).pdf

[36] Greenebaum B, Barnes F. Biological and Medical Aspects of Electromagnetic Fields. 4th ed. 2019. Available from:

https://www.taylorfrancis.com/books/edit/10.1201/9781315186641/biological-medical-aspects-electromagnetic-fields-fourth-edition-frank-barnes-ben-greenebaum

[37] Staebler P. Human Exposure to Electromagnetic Fields. Ed. Iste-Group; 2016. Available from: https://www.istegroup.com/fr/produit/exposition-humaine-aux-champs-electromagnetiques/

[38] Pennes HH. Analysis of tissue and arterial blood temperature in resting forearm. Journal of Applied Physiology. 1948;**1**:93-122

[39] Perthame B. Mathematical Models in Neuroscience. Jussieu, Paris: Pierre and Marie Curie University-Paris, J.-L. Lions Laboratory; 2023

[40] Amari S. Homogeneous nets of neuron-like elements. Biological Cybernetics. 1975;**17**:211-220

[41] Amari S-I. Dynamics of pattern formation in lateral-inhibition type neural fields. Biological Cybernetics. 1977;**27**:77-87

[42] Bressloff PC. Waves in neural media. In: From Single Neurons to Neural Fields, Lecture Notes on Mathematical Modeling in the Life Sciences. New York: Springer; 2014

[43] Brezis H. Functional Analysis, Applied Mathematics Collection for Masters. [Collection of Applied Mathematics for the Master's Degree]. Masson, Paris: Theory and Applications; 1983

[44] Françoise J-P. Oscillations in biology. In: Qualitative Analysis and Models. Vol. 46. Berlin: Mathematics & Applications, Springer-Verlag; 2005

[45] Gerstner W, Kistler WM. Spiking neuron models. In: Single Neurons, Populations, Plasticity. Cambridge: Cambridge University Press; 2002

[46] Konlack TBC, Tchuidjan R. Analysis of the Impact of Electromagnetic Waves on Humans. Cameroon: Department of Electrical and Telecommunications Engineering, University of Yaoundé; 2011

Chapter 3

# Plasmonic Excitations in Carbon Nanostructures: Carbon Nanotubes Vs. Graphene

*Pablo Martín-Luna and Javier Resta-López*


## Abstract

The interactions of charged particles with carbon nanotubes (CNTs) and graphene layers may excite plasmonic modes in the two-dimensional electron gas that constitute these shells. It has recently been proposed that plasmonic excitations could serve as an innovative approach to short-wavelength, high-gradient particle acceleration. It may transform current particle acceleration techniques and lead to new developments in high-energy physics and related disciplines. In this chapter, we compare the excited wakefields created by a charged particle traveling along a CNT and between two graphene layers. We use a theory based on a linearized hydrodynamic model in conjunction with the Poisson's equation to describe the electronic excitations produced by a fast point-like charge. In this model, surface plasmonic excitations are described by treating the electron gas as a two-dimensional plasma with additional terms in the fluid momentum equation arising from specific solid-state properties of the gas. Analytical expressions of the excited wakefields are derived that, in the case of a single-walled CNT, may be related to the resonant frequencies obtained from the dispersion relation. The dependence of the wakefields on model parameters, such as particle velocity, tube aperture, distance between graphene layers, friction parameter, and surface density, is analyzed.


**Keywords:** plasmonic excitations, wakefields, carbon nanotubes (CNTs), graphene, hydrodynamic model

## 1. Introduction

Particle accelerators significantly influence science, industry, and the economy. Traditional accelerators rely on RF technology, but their gradients are restricted to around 100 MV/m due to a phenomenon known as surface RF breakdown [1, 2]. Several alternative approaches are being explored to overcome this limitation, including dielectric laser accelerators and wakefield accelerators [3–7], as well as plasma-based accelerators [8–12]. During the 1980s and 1990s, T. Tajima and others [13–15] proposed solid-state wakefield acceleration using crystals as a potential technique to achieve acceleration gradients on the order of TV/m. In Tajima's initial proposal [13], a longitudinal wakefield is generated by high-energy ($\sim$40 keV) X-rays, which are injected into a crystal lattice at the Bragg angle, producing the Borrmann-Campbell effect [16, 17]. Similarly, ultrashort charged particle bunches can excite electric wakefields, converting the energy loss of the driving bunch into an energy gain for a properly injected witness bunch. However, the angstrom-sized channels of natural crystals limit beam intensity acceptance and increase the dechanneling rate.

In this context, carbon-based nanostructures are considered promising candidates for achieving extremely high gradients because of their exceptional thermal, mechanical, and electronic properties. For example, studies have experimentally shown the transport of 2 MeV $He^+$ ions [18] and 300 keV electrons [19] through carbon nanotubes (CNTs). As a result, CNTs can create wider channels in two dimensions and achieve longer dechanneling lengths than natural crystals [20, 21]. Therefore, carbon nanostructures, including CNTs [22–24] and graphene layers [25, 26], are currently under investigation for wakefield acceleration.

The collective oscillation of the free-electron gas (often referred to as plasmons) on carbon nanostructure surfaces (e.g., CNTs or graphene layers) can be responsible for the excitation of wakefields. This wake effect is generated due to the interaction between the driving bunch and the free electrons within these carbon nanostructures. The electronic excitations on CNTs and graphene layers have been theoretically studied employing different approaches, for example, hydrodynamic models [27–38], kinetic models [39–42], and molecular dynamics simulations [43, 44].

In this chapter, we have adopted the linearized hydrodynamic model (LHM) to derive analytical expressions of the excited wakefields in multi-walled CNTs (MWCNTs) and parallel graphene layers. In particular, a detailed study has been conducted on the wakefields excited and the stopping power induced by a proton, comparing a single-walled CNT with a radius of $a$ and two graphene sheets separated by a distance of $2a$. The LHM is chosen for its simplicity and its excellent concordance with the dielectric formalism in the random-phase approximation [28].

This chapter is organized as follows. Section 2 reviews the LHM for multi-layer CNTs and graphene layers, providing general expressions for both longitudinal and transverse wakefields excited by the interaction with a charged particle. Section 3 analyzes the dependence of the wakefields and the stopping power on model parameters, comparing a CNT with two graphene layers. Finally, a summary of the findings of this study is presented in Section 4. Atomic units are used throughout the manuscript, unless otherwise indicated.

## 2. Linearized hydrodynamic model

In this section, the LHM is used to describe excitations produced by a single charged particle on carbon nanostructure surfaces, for example, MWCNTs or graphene layers. Within this framework, the delocalized electrons of the carbon ions are considered as a free-electron gas distributed uniformly over the surface (considered infinitesimally thin), with the homogeneous density per unit area $n_0$. We consider that a driving point-like charge $Q$ is moving parallel to the $z$-axis with a constant velocity $v$, that is, with coordinates $\mathbf{r}_0(t) = (x_0, y_0, vt)$, neglecting energy losses. The homogeneous electron gas will be perturbed by the driving charge $Q$, producing a perturbation in the homogeneous electron gas on the surfaces, which can be regarded as charged fluids (confined to the surfaces) with velocity fields $\mathbf{u}_j(\mathbf{r}_j, t)$ and surface density $n(\mathbf{r}_j, t) = n_0 + n_j(\mathbf{r}_j, t)$, where $\mathbf{r}_j$ are the coordinates of a general point at the $j$th surface and $n_j(\mathbf{r}_j, t)$ is its perturbed density per unit area. In the LHM, $\mathbf{u}_j$ and $n_j$ are considered relatively small perturbations. For the analysis of wakefield dynamics, ionic motion can be neglected [45, 46], since the timescale for ionic movement is much slower than that of electrons, owing to the significantly larger mass of carbon ions compared to electrons.

In the LHM, electronic excitations on the surfaces are described by the continuity equation and the non-relativistic momentum-balance equation at each surface:

$$\frac{\partial n_j(\mathbf{r}_j,t)}{\partial t} + n_0 \nabla_j \cdot \mathbf{u}_j(\mathbf{r}_j,t) = 0, \tag{1}$$

$$\frac{\partial \mathbf{u}_j(\mathbf{r}_j,t)}{\partial t} = \nabla_j \Phi(\mathbf{r}_j,t) - \frac{\alpha}{n_0}\nabla_j n_j(\mathbf{r}_j,t) + \frac{\beta}{n_0}\nabla_j\left[\nabla_j^2 n_j(\mathbf{r}_j,t)\right] - \gamma \mathbf{u}_j(\mathbf{r}_j,t), \tag{2}$$

where we have retained only the first-order terms in $n_j$ and $\mathbf{u}_j$. In these equations, $\nabla_j$ differentiates only tangentially to the $j$th surface and $\Phi$ is the electric scalar potential. The sum of four different contributions is considered in Eq. (2). The first term on the right-hand side represents the force applied to the electrons on the $j$th surface due to the tangential component of the electric field. The total electric potential is the solution to the Poisson equation in free space and can be written as $\Phi = \Phi_0 + \Phi_{\text{ind}}$, where $\Phi_0 = \frac{Q}{|\mathbf{r}-\mathbf{r}_0|}$ is the Coulomb potential produced by the driving charge, and $\Phi_{\text{ind}}$ is the potential resulting from the perturbation of the electron fluids:

$$\Phi_{\text{ind}}(\mathbf{r},t) = -\sum_j \int \mathrm{d}^2\mathbf{r}_j \frac{n_j(\mathbf{r}_j,t)}{\|\mathbf{r}-\mathbf{r}_j\|}. \tag{3}$$

The second and third terms correspond to components of the internal interaction force within the electron gas. Specifically, the second term accounts for potential coupling with acoustic modes with $\alpha = v_F^2/2$ ($v_F = (2\pi n_0)^{1/2}$ is the Fermi velocity of the two-dimensional gas), while the third term describes single-electron excitations in the electron gas with $\beta = 1/4$. The final term represents a frictional force on the electrons resulting from scattering with the charges of the ionic lattice, where $\gamma$ denotes the friction coefficient. Additionally, it serves as a phenomenological parameter to describe the broadening of the plasmon resonance in the excitation spectra of various materials [47].

### 2.1 Carbon nanotubes

In this case, we consider an MWCNT composed of $N$ concentric cylinders with radii $a_1 < a_2 < \cdots < a_N$ [48], as illustrated in **Figure 1**.

Thus, in Eqs. (1)–(3), we use cylindrical coordinates, and, consequently, $\nabla_j = \hat{\boldsymbol{\varphi}}\frac{1}{a_j}\frac{\partial}{\partial\varphi} + \hat{\mathbf{z}}\frac{\partial}{\partial z}$, $\mathbf{r}_j = (a_j, \varphi, z)$, $\mathrm{d}^2\mathbf{r}_j = a_j \mathrm{d}\varphi \mathrm{d}z$ and the position of the driving particle is $\mathbf{r}_0(t) = (r_0, \varphi_0, vt)$. Eqs. (1)–(3) can be solved by performing a Fourier-Bessel (FB) transform, whose definition for an arbitrary function $f$ is:

$$f(\varphi,z,t) = \sum_{m=-\infty}^{\infty} \int_{-\infty}^{\infty} \frac{\mathrm{d}k}{(2\pi)^2} \int_{-\infty}^{\infty} \frac{\mathrm{d}\omega}{2\pi} \mathrm{e}^{ikz+im\varphi-i\omega t}\tilde{f}(m,k,\omega). \tag{4}$$

The FB transform of the Coulomb potential created by the driving point-like charge and the induced potential are given, respectively, by [24]:

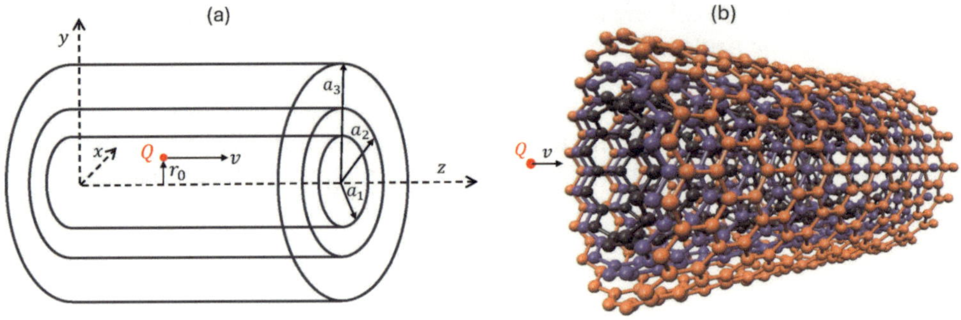


**Figure 1.**
*(a) Schematic of a charge Q moving parallel to the z-axis inside an MWCNT with N = 3 cylinders. (b) Schematic of the hexagonal lattice of a triple-walled CNT.*

$$\tilde{\Phi}_0(r, m, k, \omega) = 2\pi Q g(r, r_0; m, k)\delta(\omega - kv)e^{-im\varphi_0}, \tag{5}$$

$$\tilde{\Phi}_{\text{ind}}(r, m, k, \omega) = -\sum_{j=1}^{N} g(r, a_j; m, k) a_j \tilde{n}_j(m, k, \omega), \tag{6}$$

where $\delta$ is the Dirac delta, $g(r, r'; m, k) \equiv 4\pi I_m(k|r_{\min}|)K_m(k|r_{\max}|)$ with $r_{\min} = \min(r, r')$, $r_{\max} = \max(r, r')$ and $I_m(x)$ and $K_m(x)$ are the modified Bessel functions of integer order $m$.

After eliminating $\mathbf{u}_j$ in Eq. (2) by using the continuity equation and applying the FB definition, the FB transform of the perturbed densities can be calculated by solving the system of $N$ coupled linear equations:

$$S_j(m, k, \omega)\tilde{n}_j(m, k, \omega) - \sum_{l=1}^{N} G_{jl}(m, k)\tilde{n}_l(m, k, \omega) = B_j(m, k, \omega), \tag{7}$$

where the following functions have been defined:

$$S_j(m, k, \omega) = \omega(\omega + i\gamma) - \alpha\left(k^2 + m^2/a_j^2\right) - \beta\left(k^2 + m^2/a_j^2\right)^2, \tag{8}$$

$$G_{jl}(m, k) = n_0 a_l \left(k^2 + m^2/a_j^2\right) g(a_j, a_l; m, k), \tag{9}$$

$$B_j(m, k, \omega) = -n_0\left(k^2 + m^2/a_j^2\right)\tilde{\Phi}_0(a_j, m, k, \omega). \tag{10}$$

It is worth noting that Eq. (7) can be expressed as the following matrix equation:

$$M\vec{\tilde{n}} = \vec{B}, \qquad M_{jj} = S_j - G_{jj}, \quad M_{ij} = -G_{ij} \quad (i \neq j). \tag{11}$$

Therefore, the resonant frequencies of the collective excitations in the electron fluids can be obtained by solving the eigenvalue equation $\det(M) = 0$ for $\gamma = 0$. Once the FB of the perturbed densities has been obtained, the induced potential can be calculated as:

$$\Phi_{\text{ind}}(\mathbf{r}, t) = \frac{2}{(2\pi)^3} \sum_{m=-\infty}^{+\infty} e^{im\varphi} \int_0^{+\infty} \text{Re}\left[\tilde{\Phi}_{\text{ind}}(r, m, k, kv)e^{ik\zeta}\right] dk, \tag{12}$$

where $\tilde{\Phi}_{\text{ind}}$ can be obtained using Eq. (6) and the comoving coordinate $\zeta = z - vt$ has been defined. The induced longitudinal and transverse wakefields can be determined by calculating the corresponding partial derivatives: $W_z = -\partial_z \Phi_{\text{ind}}$ and $W_r = -\partial_r \Phi_{\text{ind}}$.

If we consider a single-walled CNT with radius $a \equiv a_1$, the FB transform of the perturbed density can be obtained from Eq. (7) as:

$$\tilde{n}_1(m,k,\omega) = \frac{-2\pi Q n_0 (k^2 + m^2/a^2) g(a, r_0; m, k) \delta(\omega - kv) e^{-im\varphi_0}}{D_m(k,\omega)}, \tag{13}$$

with $D_m(k,\omega) = \omega(\omega + i\gamma) - \omega_m^2(k)$, where

$$\omega_m^2(k) = \alpha(k^2 + m^2/a^2) + \beta(k^2 + m^2/a^2)^2 + \Omega_p^2 a^2 (k^2 + m^2/a^2) K_m(k|a|) I_m(k|a|) \tag{14}$$

are the plasmon dispersion curves and $\Omega_p = \sqrt{4\pi n_0/a}$ is the plasma frequency. The FB transform of the electric potential is $\tilde{\Phi}_{\text{ind}}(r,m,k,\omega) = -g(r,a;m,k) a \tilde{n}_1(m,k,\omega)$. As Eq. (5), implies $\omega = kv$, the condition for plasma resonance is obtained when $k_m v = \omega_m(k_m)$ (i.e., when $D_m(k_m, k_m v) = 0$ for $\gamma \to 0^+$). It is worth noting that the wakefields can be approximately calculated analytically in terms of the roots $k_m$ if $\gamma \to 0^+$ [23, 48].

### 2.2 Graphene layers

In this case, we consider $N$ parallel graphene layers at planes $y_1 < y_2 < \cdots < y_N$ [26], as illustrated in **Figure 2**.

Hence, in Eqs. (1)–(3), we use Cartesian coordinates and, consequently, $\nabla_j = \hat{\mathbf{x}} \frac{\partial}{\partial x} + \hat{\mathbf{z}} \frac{\partial}{\partial z}$, $\mathbf{r}_j = \left(x, y_j, z\right)$, and $d^2\mathbf{r}_j = dxdz$. The LHM can be solved similarly to the case of MWCNTs, although in the present case we have to use a Fourier transform with respect to coordinates in the $xz$-plane, $\mathbf{R} = (x,z) \to \mathbf{k} = (k_x, k_z)$, and time, $t \to \omega$, which is defined by

$$A(\mathbf{R}, y, t) = \frac{1}{(2\pi)^3} \int \tilde{A}(\mathbf{k}, y, \omega) e^{-i(\omega t - \mathbf{k} \cdot \mathbf{R})} d^2\mathbf{k} d\omega. \tag{15}$$

The Fourier transform of the electric potential created by the driving charge $\tilde{\Phi}_0$ and the induced electric potential created by the graphene layers $\tilde{\Phi}_{\text{ind}}$ are, respectively,

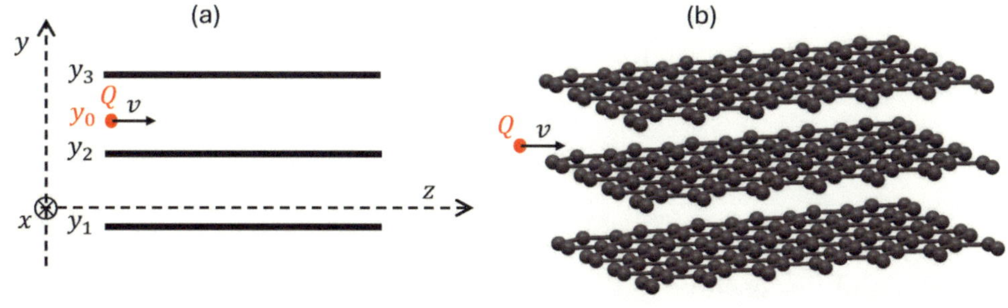


**Figure 2.**
*(a) Schematic of a charge Q moving parallel to the z-axis in a system with N = 3 graphene layers. (b) Schematic of the hexagonal lattice of the graphene layers.*

$$\tilde{\Phi}_0(\mathbf{k}, y, \omega) = \frac{(2\pi)^2 Q\delta(\omega - \mathbf{k}\cdot\mathbf{v})}{k} e^{-k|y-y_0|}, \tag{16}$$

$$\tilde{\Phi}_{\text{ind}}(\mathbf{k}, y, \omega) = -\sum_{j=1}^{N} \frac{2\pi}{k} \tilde{n}_j(\mathbf{k}, \omega) e^{-k|y-y_j|}, \tag{17}$$

where we have defined $k = \sqrt{k_x^2 + k_z^2}$. If the continuity equation is utilized to eliminate $\mathbf{u}_j$ in Eq. (2) and the Fourier transform definition is applied, Eq. (15), the Fourier transform of the perturbed densities is related by the system of $N$ coupled linear equations:

$$S_j(\mathbf{k}, \omega)\tilde{n}_j(\mathbf{k}, \omega) - \sum_{l=1}^{N} G_{jl}(\mathbf{k})\tilde{n}_l(\mathbf{k}, \omega) = B_j(\mathbf{k}, \omega), \tag{18}$$

where the following functions are defined:

$$S_j(\mathbf{k}, \omega) = \omega(\omega + i\gamma) - \alpha k^2 - \beta k^4, \tag{19}$$

$$G_{jl}(\mathbf{k}) = 2\pi n_0 k e^{-k|y_j - y_l|}, \tag{20}$$

$$B_j(\mathbf{k}, \omega) = -n_0 k^2 \tilde{\Phi}_0\left(\mathbf{k}, y_j, \omega\right). \tag{21}$$

It is important to note the similarity with the expressions obtained in the MWCNTs case: Eqs. (7)–(10). In particular, Eq. (18) can be expressed as the matrix Eq. (11), and, consequently, the dispersion relations may be obtained by solving the eigenvalue equation $\det(M) = 0$ for $\gamma = 0$. Once the $\tilde{n}_j$ have been obtained by solving Eq. (18), the perturbed densities and the electric potential can be determined by utilizing the definition of the Fourier transform, Eq. (15). In particular, by leveraging the symmetry properties of the real and imaginary parts of $\tilde{n}_j$ in $\mathbf{k}$, the perturbed densities and the induced potential can be expressed as:

$$n_j(\mathbf{R}, y, t) = \frac{4}{(2\pi)^3} \int_0^\infty \int_0^\infty \mathrm{Re}\left[\tilde{n}_j(\mathbf{k}, y, k_z v) e^{ik_z\zeta}\right] \cos(k_x x) dk_x dk_z, \tag{22}$$

$$\Phi_{\text{ind}}(\mathbf{R}, y, t) = \frac{4}{(2\pi)^3} \int_0^\infty \int_0^\infty \mathrm{Re}\left[\tilde{\Phi}_{\text{ind}}(\mathbf{k}, y, k_z v) e^{ik_z\zeta}\right] \cos(k_x x) dk_x dk_z, \tag{23}$$

where the comoving coordinate $\zeta = z - vt$ has been defined, as in the case of an MWCNT. Finally, the induced wakefields can be determined by calculating the corresponding partial derivatives:

$$W_x(\mathbf{R}, y, t) = -\frac{\partial \Phi_{\text{ind}}}{\partial x}, \quad W_y(\mathbf{R}, y, t) = -\frac{\partial \Phi_{\text{ind}}}{\partial y}, \quad W_z(\mathbf{R}, y, t) = -\frac{\partial \Phi_{\text{ind}}}{\partial z}. \tag{24}$$

## 3. Results and discussion

The objective of this research is to compare the wakefields excited inside a CNT with radius $a$ with those excited between two graphene sheets separated a distance $2a$,

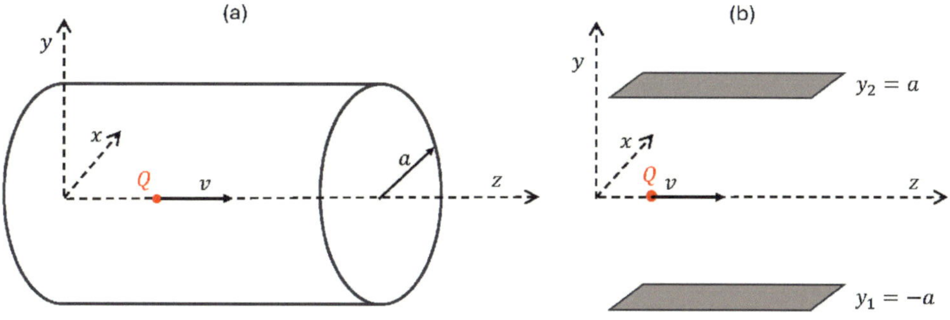


**Figure 3.**
*Schematic of a charge Q moving along the z-axis in (a) a CNT with radius a and (b) two graphene layers at planes $y_1 = -a$ and $y_2 = a$.*

that is, the geometries depicted in **Figure 3**. Unless otherwise indicated, it is assumed that the surface electron densities can be approximated by the electron-gas density of a graphite sheet: $n_0 = n_g = 1.53 \times 10^{20} \mathrm{m}^{-2}$ [28, 49].

The roots $k_m$ provided by the plasma resonance are crucial in describing the behavior of the wakefields in a CNT. For this reason, this section begins with a detailed analysis of the dispersion relation of a CNT, comparing it with the dispersion relations of two graphene layers.

### 3.1 Dispersion relation

**Figure 4** shows the dispersion curves $\omega_m(k)$, Eq. (14), for several angular-momentum modes at different CNT radii: (a) $a = 1$ nm and (b) $a = 2$ nm. The resonant wavenumbers $k_m$ occur where the $kv$ line intersects with the dispersion relation $\omega_m(k)$. Consequently, as the driving velocity increases, the resonant wavenumber $k_m$ (if it exists) decreases. The fundamental mode $m = 0$ is the only one that may not satisfy the plasma resonance condition at sufficiently high velocities. From Eq. (14), it is evident that plasma resonance cannot be achieved if $v < \sqrt{\alpha} = v_F/\sqrt{2}$ and that the dispersion curves increase if the surface density $n_0$ increases. For completeness, the dispersion curves for two graphene layers separated a distance of $2a$

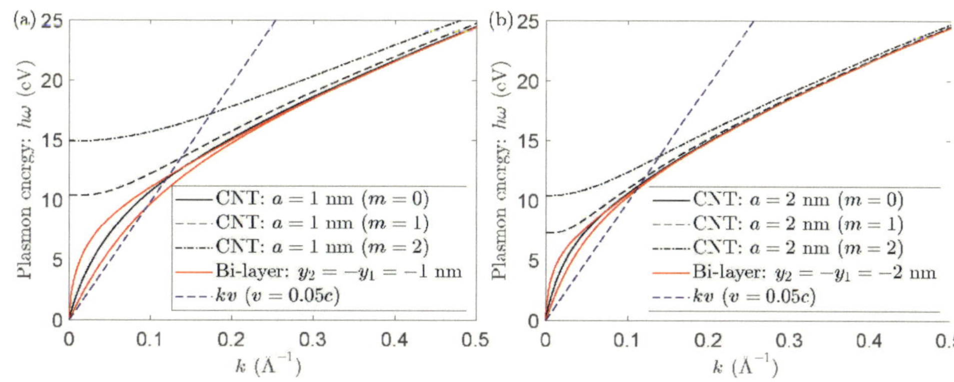


**Figure 4.**
*Plasmon dispersion relations $\omega_m(k)$ for different modes m for a CNT with (a) $a = 1$ nm and (b) $a = 2$ nm. The resonances $k_m$ are the intersection of a kv line (plotted for $v = 0.05c$) with the dispersion curves $\omega_m(k)$. For comparison, we include the dispersion relations for two graphene layers separated by a distance of 2a.*

are also included, which are given by $\omega_\pm(k) = \sqrt{\omega_1^2 \pm \Delta}$, where $\omega_1^2 = \alpha k^2 + \beta k^4 + 2\pi n_0 k$ is the dispersion relation of a single graphene layer and $\Delta = 2\pi n_0 k e^{-k|y_1 - y_2|}$ describes the electrostatic interaction between both layers. It can be observed that they resemble the fundamental mode $\omega_0(k)$ of the CNT, especially as $a$ increases because the interaction term $\Delta$ decreases exponentially with the separation between layers.

### 3.2 Excited wakefields

For the sake of simplicity, in the following calculations, it is considered that the point-like charged particle is a proton (i.e., $Q = 1$) traveling along the $z$-axis. Consequently, the fundamental mode $\omega_0(k)$ is the only mode that contributes to the wakefield excitation in the CNT. The integrals in Eqs. (22) and (23) require a finite value of $\gamma$ for proper convergence. Therefore, a value of $\gamma = 10^{-3}$ a.u.$\approx 4.13 \times 10^{13}$s$^{-1}$ will be used in all cases.

**Figure 5** compares the longitudinal wakefield $W_z$ and the transverse wakefield $W_y$ in the $y\zeta$-plane in a CNT and two graphene layers, respectively. It can be seen that the wakefields are similar, although the wakefield intensity is lower in the graphene layers. Additionally, the wakefields diminish rapidly behind the driving particle in the case of the graphene layers, unlike in the case of the CNT, where they hardly decrease. The friction coefficient $\gamma$ causes an exponential decay of the wakefields behind the driving particle, but a sufficiently small value has been chosen so that it is not noticeable in these cases. The reason for the rapid decay in the graphene layers is that the wakefields spread out in the plane of the layers (i.e., the $xz$-plane), whereas in the case of the CNT, the wakefields are confined within the tube. To observe this, **Figure 6**

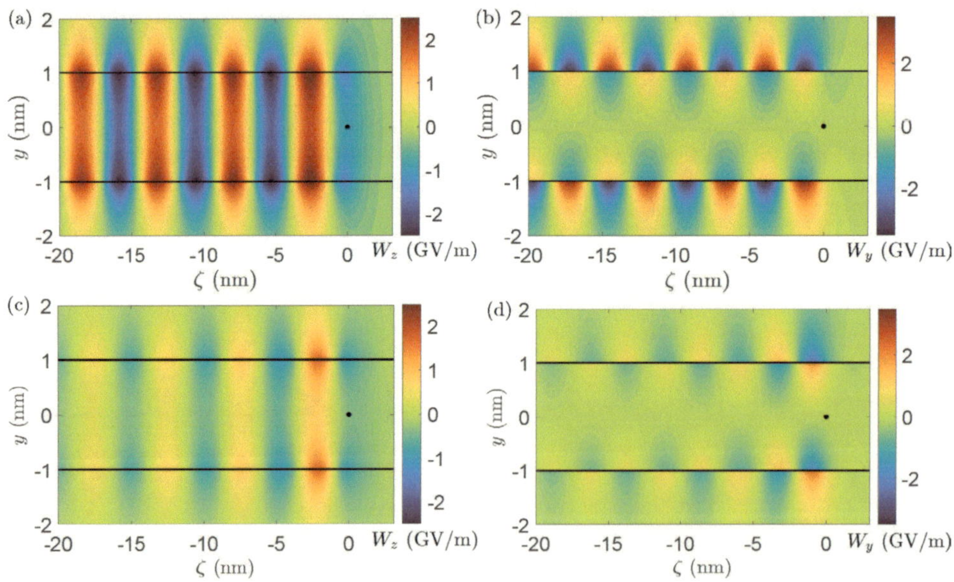


**Figure 5.**
*Induced wakefields (a) $W_z$ and (b) $W_y$ in the $y\zeta$-plane for a proton traveling on the z-axis with a velocity $v = 0.05c$ within a CNT with radius $a = 1$ nm. In (c) and (d), we consider two graphene layers located at $y_2 = -y_1 = a$. In both cases, the friction parameter is $\gamma = 10^{-3}$ a.u.$\approx 4.13 \times 10^{13}s^{-1}$. The wakefield $W_x$ is zero in the $y\zeta$-plane. The proton is indicated with a black point and the CNT wall and the graphene layers with black lines.*

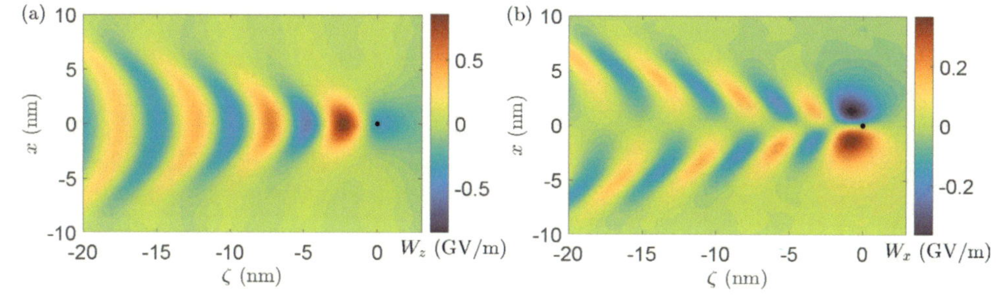


**Figure 6.**
*Induced wakefields (a) $W_z$ and (b) $W_x$ in the $x\zeta$-plane for a proton traveling on the z-axis with a velocity $v = 0.05c$ within two graphene layers located at $y_2 = -y_1 = a$. The wakefield $W_y$ is zero in the $x\zeta$-plane. The proton is indicated with a black point.*

depicts the wakefields $W_z$ and $W_x$ in the $x\zeta$-plane in the bi-layer configuration. Due to the cylindrical symmetry of the CNT, the wakefields $W_z$ and $W_x$ in the $x\zeta$-plane are identical to the wakefields $W_z$ and $W_y$, respectively, in the $y\zeta$-plane, that is, the wakefields depicted in **Figure 5(a)** and **(b)**. This difference in wakefields is highly significant for particle acceleration purposes. In CNTs, there are regions where a witness beam experiences both acceleration and focusing simultaneously. However, in the bi-layer configuration, if acceleration and focusing are achieved in the $y\zeta$-plane ($x\zeta$-plane), then there would be defocusing in the $x\zeta$-plane ($y\zeta$-plane).

For completeness, **Figure 7** shows the perturbed density $n_1/n_0$ in the CNT and the graphene layer. Due to the symmetry, the perturbed density depends only on $\zeta$ in the CNT and $n_1 = n_2$ in the bi-layer configuration. Similar to the wakefields in the $x\zeta$-plane, the perturbed density spreads out in the graphene layers. It is worth noting that the perturbed densities satisfy $n_1 \ll n_0$, as assumed in the linearized hydrodynamic model.

**Figure 8** shows the induced longitudinal wakefield $W_z$ in the $z$-axis in order to study the effect of each parameter: the driving velocity $v$, the CNT radius $a$ (separation between layers $2a$), and the surface density $n_0$. It can be observed that the wavelength of the wakefield increases with the velocity, whereas the intensity of the wakefield decreases for low velocities and for high velocities, that is, there is an optimal velocity for which the wakefield intensity is maximum. Moreover, the wavelength of the plasmonic excitations increases with the velocity $v$. As can be seen in

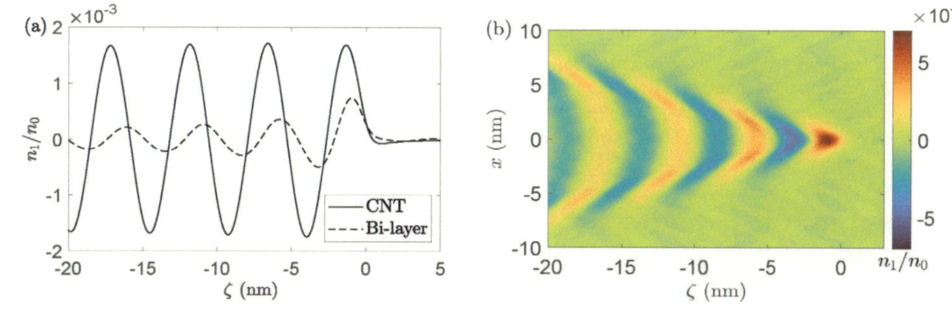


**Figure 7.**
*(a) Normalized perturbed density $n_1/n_0$ for a proton traveling on the z-axis with a velocity $v = 0.05c$ within a CNT with radius $a = 1$ nm and two graphene layers located at $y_2 = -y_1 = a$. In (b) we show the perturbed density in the layer plane, that is, at $y_1 = -a$. Due to symmetry, the perturbed density in both layers is identical.*

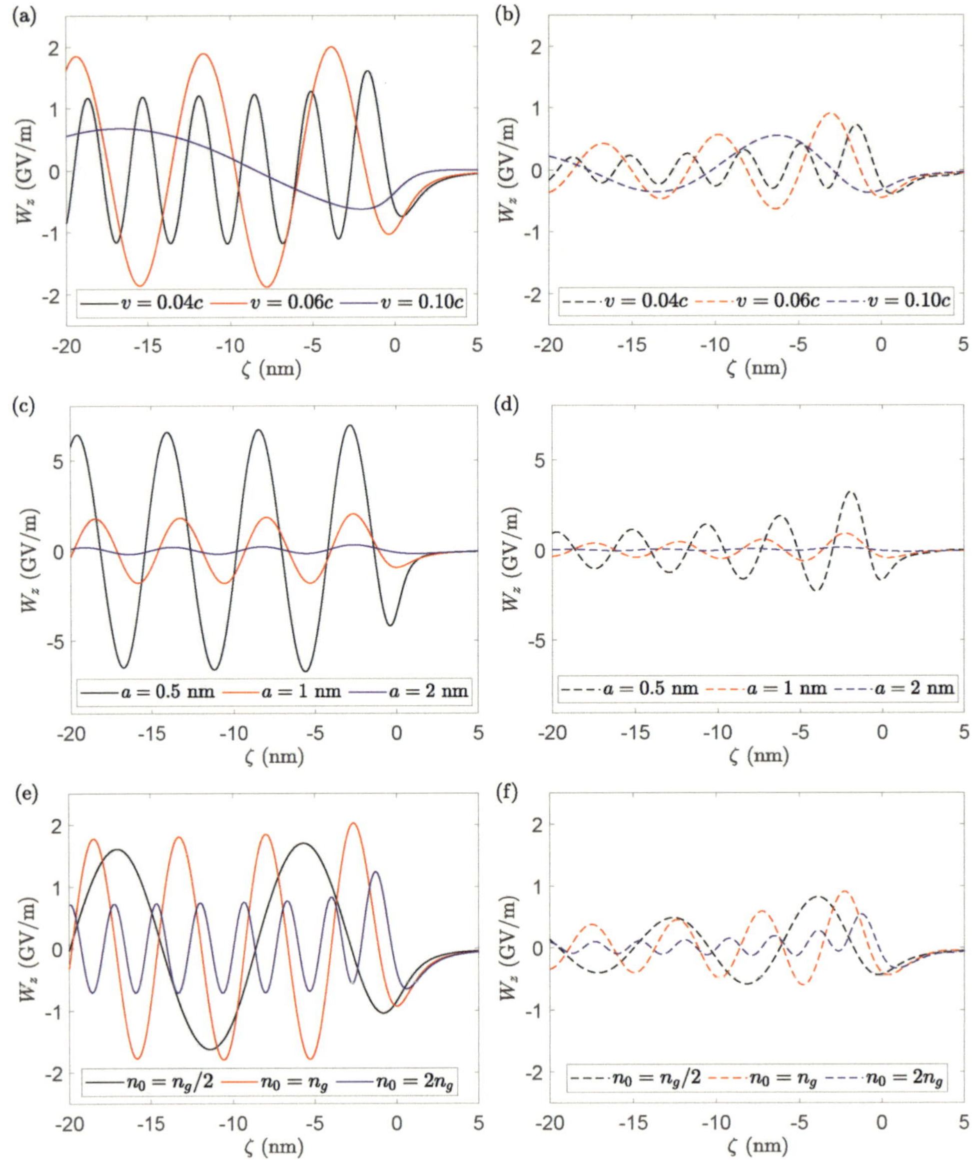


**Figure 8.**
*Longitudinal wakefield $W_z$ along the z-axis for a proton traveling on the z-axis for different (a)-(b) velocities $v$, (c)-(d) CNT radii $a$ or separation between layers $2a$ and (e)-(f) surface density $n_0$. In all cases, unless otherwise indicated in the corresponding legend, the following parameters are used: $v = 0.05c$, $a = 1$ nm, $n_0 = n_g$. Solid lines indicate the CNT case (left column) and dashed lines the bi-layer configuration (right column).*

**Figure 8(c)** and **(d)**, the intensity of the wakefields increases as $a$ diminishes (this behavior is consistently observed in the graphene layers, but for CNTs the intensity decreases for very small radii; see **Figure 11**). Finally, **Figure 8(e)** and **(f)** illustrate that the intensity of the wakefields decreases for low and high surface densities, that is, there is an optimum density (in this case, $n_0 \approx n_g$). The wavelength of the wakefields decreases with the surface density. The dependence of the wavelength on velocity and surface density can be directly deduced from the dispersion relation. As the velocity decreases (or surface density increases), the value of the resonant wavenumber $k_0$ increases, leading to a decrease in wavelength ($\lambda \sim 2\pi/k_0$). Finally, it is observed that, in general, CNTs produce more intense wakefields that do not decay rapidly and have longer wavelengths compared to graphene layers.

### 3.3 Maximum wakefield and stopping power

This section is dedicated to comparing the maximum longitudinal wakefield excited along the $z$-axis $\left(W_z^{max}\right)$ in the CNT and in the bi-layer configuration. For completeness, we will examine the stopping power, which refers to the energy loss of a channeled particle per unit path length caused by the collective electron excitations. This magnitude can be calculated as

$$S = -QW_z|_{\mathbf{r}=\mathbf{r}_0}, \tag{25}$$

and exhibits a behavior similar to that of the excited longitudinal wakefield amplitude [24]. **Figure 9** depicts $W_z^{max}$ and the stopping power as a function of the driving velocity for different values of $a$. Both the CNT and the bi-layer configuration exhibit qualitatively similar behavior: $W_z^{max}$ decreases as the parameter $a$ increases. In both cases, the maximum is reached at a similar velocity (slightly lower in the CNT). In the CNT, a larger wakefield (approximately twice as large) can be obtained, but it decreases more rapidly at high velocities compared to the graphene layers. It is also observed that $W_z^{max}$ and the stopping power have a very similar trend, except at very low velocities where $W_z^{max}$ tends to a finite value and the stopping power to zero. In fact, at these very low velocities, no plasmonic excitation (i.e., an oscillating wakefield) occurs, and the wakefield is only generated near the driving particle [35, 36].

**Figure 10** shows $W_z^{max}$ and the stopping power as a function of the driving velocity for different surface densities $n_0$. Again, both configurations (CNT and graphene layers) exhibit qualitatively similar behavior. It is interesting to note that the maximum of $W_z^{max}$ and the stopping power shift to higher velocities with the surface density, although their peak values remain approximately constant.

Finally, **Figure 11** analyzes $W_z^{max}$ and the stopping power as a function of $a$. It is observed that in graphene layers, these magnitudes increase as the layers are closer together, whereas for CNTs, it exists an optimal radius.

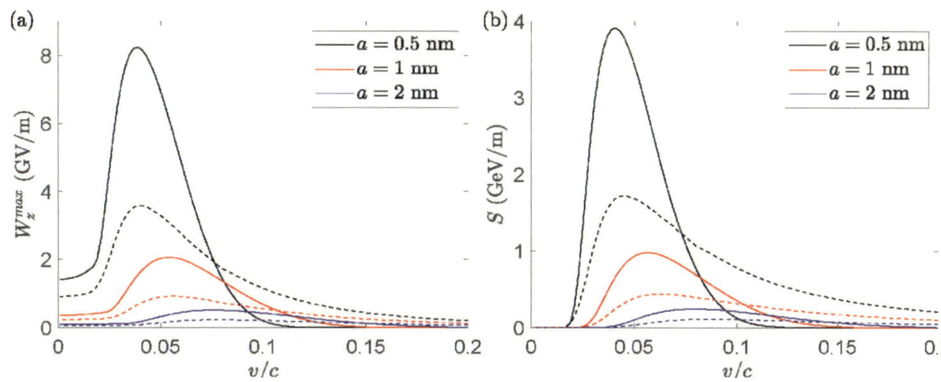


**Figure 9.**
*(a) Maximum longitudinal wakefield $W_z^{max}$ and (b) stopping power as a function of the driving velocity for different values of a. Solid lines indicate the CNT case and dashed lines the bi-layer configuration.*

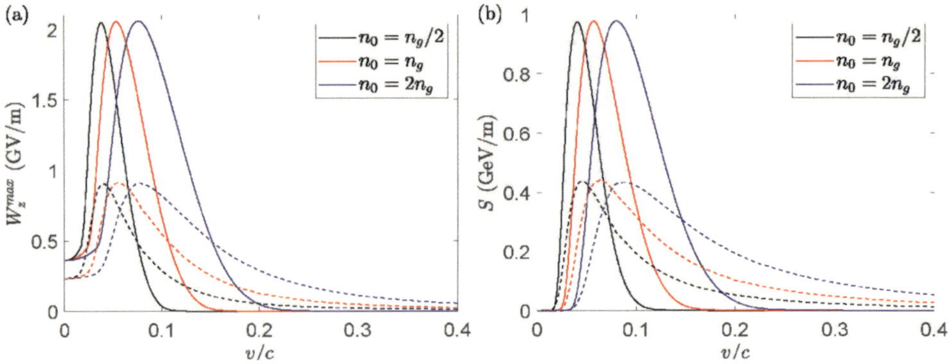


**Figure 10.**
*(a) Maximum longitudinal wakefield $W_z^{max}$ and (b) stopping power as a function of the driving velocity for different values of the surface density $n_0$ for $a = 1$ nm. Solid lines indicate the CNT case and dashed lines the bi-layer configuration.*

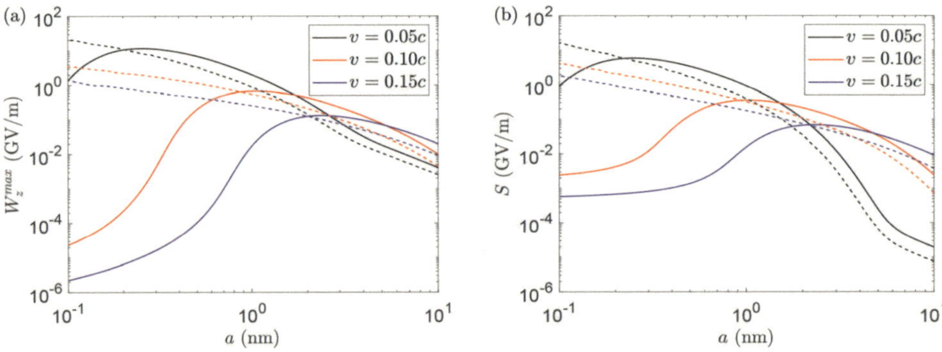


**Figure 11.**
*(a) Maximum longitudinal wakefield $W_z^{max}$ and (b) stopping power as a function of the CNT radius a (layers separated a distance 2a) for different values of the driving velocity. Solid lines indicate the CNT case and dashed lines the bi-layer configuration.*

## 4. Conclusions and outlook

The linear hydrodynamic model has been revisited to derive general expressions for the electric wakefields excited by a point-like charge in carbon nanostructures, in particular MWCNT and parallel graphene layers. The excitation in a CNT with radius $a$ has been compared in detail with that between two graphene layers separated by a distance of $2a$.

As the wakefields of a CNT are closely related to the dispersion relation, the dispersion curves have been initially plotted. It has been observed that the bi-layer configuration exhibits two dispersion curves that resemble the dispersion relation $\omega_0(k)$ of the CNT, especially as $a$ increases.

Subsequently, the excited wakefields have been compared in a specific case. It has been observed that in the plane perpendicular to the graphene layers (i.e., the $yz$-plane), the fields qualitatively resemble each other, except that the wakefields decay with distance in the bi-layer case. The rapid decay in the graphene layers occurs because the wakefields spread out in the plane of the layers (i.e., the $xz$-plane), whereas in the case of the CNT, the wakefields are confined within the tube structure.

Thus, a witness beam could experience acceleration and focusing in a CNT, whereas in the graphene layers, it would be focused in one plane and defocused in the perpendicular plane.

Finally, the maximum excited longitudinal wakefield along the $z$-axis and the stopping power have been studied. These magnitudes exhibit qualitatively similar behavior in both the CNT and the graphene layers, with the exception that in the CNT there is an optimal radius (i.e., a radius that maximizes these magnitudes), whereas in the graphene layers, these magnitudes increase as the separation decreases.

In summary, the excitation of very intense wakefields in carbon nanostructures has been demonstrated, which could pave the way for new applications in particle acceleration or radiation sources.

## Acknowledgements

This work has been supported by Ministerio de Universidades (Gobierno de España) under grant agreement FPU20/04958 and the Generalitat Valenciana under grant agreement CIDEGENT/2019/058.

## Abbreviations

| | |
|---|---|
| CNT | carbon nanotube |
| FB | Fourier-Bessel |
| LHM | linearized hydrodynamic model |
| MWCNT | multi-walled carbon nanotube |

## Author details

Pablo Martín-Luna[1*†] and Javier Resta-López[2*†]

1 Instituto de Física Corpuscular (IFIC), Universitat de València - Consejo Superior de Investigaciones Científicas, Paterna, Spain

2 Instituto de Ciencia de los Materiales (ICMUV), Universidad de Valencia, Valencia, Spain

*Address all correspondence to: pablo.martin@uv.es and javier2.resta@uv.es

† These authors contributed equally.

**IntechOpen**

© 2025 The Author(s). Licensee IntechOpen. This chapter is distributed under the terms of the Creative Commons Attribution License (http://creativecommons.org/licenses/by/4.0), which permits unrestricted use, distribution, and reproduction in any medium, provided the original work is properly cited.

## References

[1] Abe T, Kageyama T, Sakai H, Takeuchi Y, Yoshino K. Breakdown study based on direct in situ observation of inner surfaces of an RF accelerating cavity during a high-gradient test. Physical Review Accelerators and Beams. 2016;**19**:102001. DOI: 10.1103/PhysRevAccelBeams.19.102001

[2] Shao J. Investigations on rf Breakdown Phenomenon in High Gradient Accelerating Structures. Singapore: Springer; 2018. DOI: 10.1007/978-981-10-7926-9

[3] Thompson MC, Badakov H, Cook AM, Rosenzweig JB, Tikhoplav R, Travish G, et al. Breakdown limits on Gigavolt-per-meter electron-beam-driven Wakefields in dielectric structures. Physical Review Letters. 2008;**100**:214801. DOI: 10.1103/PhysRevLett.100.214801

[4] Breuer J, Hommelhoff P. Laser-based acceleration of nonrelativistic electrons at a dielectric structure. Physical Review Letters. 2013;**111**:134803. DOI: 10.1103/PhysRevLett.111.134803

[5] Peralta E, Soong K, England R, Colby E, Wu Z, Montazeri B, et al. Demonstration of electron acceleration in a laser-driven dielectric microstructure. Nature. 2013;**503**(7474):91-94. DOI: 10.1038/nature12664

[6] England RJ, Noble RJ, Bane K, Dowell DH, Ng CK, Spencer JE, et al. Dielectric laser accelerators. Reviews of Modern Physics. 2014;**86**:1337-1389. DOI: 10.1103/RevModPhys.86.1337

[7] O'Shea B, Andonian G, Barber S, Fitzmorris K, Hakimi S, Harrison J, et al. Observation of acceleration and deceleration in gigaelectron-volt-per-metre gradient dielectric wakefield accelerators. Nature Communications. 2016;**7**(1):1-7. DOI: 10.1038/ncomms12763

[8] Tajima T, Dawson JM. Laser electron accelerator. Physical Review Letters. 1979;**43**:267-270. DOI: 10.1103/PhysRevLett.43.267

[9] Chen P, Dawson JM, Huff RW, Katsouleas T. Acceleration of electrons by the interaction of a bunched electron beam with a plasma. Physical Review Letters. 1985;**54**:693-696. DOI: 10.1103/PhysRevLett.54.693

[10] Gordon D, Tzeng KC, Clayton CE, Dangor AE, Malka V, Marsh KA, et al. Observation of electron energies beyond the linear dephasing limit from a laser-excited relativistic plasma wave. Physical Review Letters. 1998;**80**:2133-2136. DOI: 10.1103/PhysRevLett.80.2133

[11] Leemans WP, Nagler B, Gonsalves AJ, Tóth C, Nakamura K, Geddes CG, et al. GeV electron beams from a centimetre-scale accelerator. Nature Physics. 2006;**2**(10):696-699. DOI: 10.1038/nphys418

[12] Litos M, Adli E, An W, Clarke C, Clayton C, Corde S, et al. High-efficiency acceleration of an electron beam in a plasma wakefield accelerator. Nature. 2014;**515**(7525):92-95. DOI: 10.1038/nature13882

[13] Tajima T, Cavenago M. Crystal x-ray accelerator. Physical Review Letters. 1987;**59**:1440-1443. DOI: 10.1103/PhysRevLett.59.1440

[14] Chen P, Noble RJ. A solid state accelerator. AIP Conference Proceedings. 1987;**156**:222-227. DOI: 10.1063/1.36458

[15] Chen P, Noble RJ. Crystal channel collider: Ultra-high energy and luminosity in the next century. AIP Conference Proceedings. 1997;**398**: 273-285. DOI: 10.1063/1.53055

[16] Borrmann G. Die absorption von röntgenstrahlen im fall der interferenz. Zeitschrift für Physik. 1950;**127**(4): 297-323. DOI: 10.1007/BF01329828

[17] Campbell H. X-ray absorption in a crystal set at the Bragg angle. Journal of Applied Physics. 1951;**22**(9):1139-1142. DOI: 10.1063/1.1700122

[18] Zhu Z, Zhu D, Lu R, Xu Z, Zhang W, Xia H. The experimental progress in studying of channeling of charged particles along nanostructure. In: Dabagov SB, editor. International Conference on Charged and Neutral Particles Channeling Phenomena, Society of Photo-Optical Instrumentation Engineers (SPIE) Conference Series. Vol. 5974. Bellingham, WA: SPIE; 2006. pp. 382-389. DOI: 10.1117/12.640101

[19] Chai G, Heinrich H, Chow L, Schenkel T. Electron transport through single carbon nanotubes. Applied Physics Letters. 2007;**91**(10):103101. DOI: 10.1063/1.2778551

[20] Biryukov VM, Bellucci S. Nanotube diameter optimal for channeling of high-energy particle beam. Physics Letters B. 2002;**542**(1):111-115. DOI: 10.1016/S0370-2693(02)02276-1

[21] Bellucci S, Biryukov VM, Cordelli A. Channeling of high-energy particles in a multi-wall nanotube. Physics Letters B. 2005;**608**(1):53-58. DOI: 10.1016/j.physletb.2005.01.003

[22] Bonatto A, Xia G, Apsimon O, Bontoiu C, Kukstas E, Rodin V, et al. Exploring ultra-high-intensity wakefields in carbon nanotube arrays: An effective plasma-density approach. Physics of Plasmas. 2023;**30**(3):033105. DOI: 10.1063/5.0134960

[23] Martín-Luna P, Bonatto A, Bontoiu C, Xia G, Resta-López J. Excitation of wakefields in carbon nanotubes: A hydrodynamic model approach. New Journal of Physics. 2023;**25**(12):123029. DOI: 10.1088/1367-2630/ad127c

[24] Martín-Luna P, Bonatto A, Bontoiu C, Xia G, Resta-López J. Plasmonic excitations in double-walled carbon nanotubes. Results in Physics. 2024;**60**:107698. DOI: 10.1016/j.rinp.2024.107698

[25] Bonțoiu C, Apsimon O, Kukstas E, Rodin V, Yadav M, Welsch C, et al. TeV/m catapult acceleration of electrons in graphene layers. Scientific Reports. 2023; **13**:1330. DOI: 10.1038/s41598-023-28617-w

[26] Martín-Luna P, Bonatto A, Bontoiu C, Lei B, Xia G, Resta-López J. Plasmonic excitations in graphene layers. arXiv. 2025. Available from: https://arxiv.org/abs/2501.05998

[27] Stöckli T, Bonard JM, Châtelain A, Wang ZL, Stadelmann P. Collective oscillations in a single-wall carbon nanotube excited by fast electrons. Physical Review B. 2001;**64**:115424. DOI: 10.1103/PhysRevB.64.115424

[28] Wang YN, Mišković ZL. Interactions of fast ions with carbon nanotubes: Self-energy and stopping power. Physical Review A. 2004;**69**:022901. DOI: 10.1103/PhysRevA.69.022901

[29] Mowbray DJ, Mišković ZL, Goodman FO, Wang YN. Interactions of fast ions with carbon nanotubes: Two-fluid model. Physical Review B. 2004;**70**:

195418. DOI: 10.1103/PhysRevB.70.195418

[30] Mowbray DJ, Mišković ZL, Goodman FO, Wang YN. Wake effect in interactions of fast ions with carbon nanotubes. Physics Letters A. 2004; **329**(1):94-99. DOI: 10.1016/j.physleta.2004.06.090

[31] Mowbray DJ, Chung S, Mišković ZL, Goodman FO, Wang YN. Dynamic interactions of fast ions with carbon nanotubes. Nuclear instruments and methods in physics research section B: Beam interactions with materials and atoms. Atomic Collisions in Solids. 2005; **230**(1):142-147. DOI: 10.1016/j.nimb.2004.12.032

[32] Chung S, Mowbray DJ, Mišković ZL, Goodman FO, Wang YN. Dynamic interactions of fast ions with multiwalled carbon nanotubes. Radiation Physics and Chemistry. 2007;**76**(3):524-528. DOI: 10.1016/j.radphyschem.2005.09.020

[33] Wei L, Wang YN. Quantum ion-acoustic waves in single-walled carbon nanotubes studied with a quantum hydrodynamic model. Physical Review B. 2007;**75**:193407. DOI: 10.1103/PhysRevB.75.193407

[34] Radović I, Hadžievski L, Bibić N, Mišković ZL. Dynamic-polarization forces on fast ions and molecules moving over supported graphene. Physical Review A. 2007;**76**(4):042901. DOI: 10.1103/PhysRevA.76.042901

[35] Radović I, Borka D. Interactions of fast charged particles with supported two-dimensional electron gas: One-fluid model. Physics Letters A. 2010;**374**(13): 1527-1533. DOI: 10.1016/j.physleta.2010.01.056

[36] Radović I, Borka D. Wake effect in interactions of fast ions with supported two-dimensional electron gas. Nuclear Instruments and Methods in Physics Research Section B: Beam Interactions with Materials and Atoms. 2010;**268**(17): 2649-2654. DOI: 10.1016/j.nimb.2010.06.043

[37] Li CZ, Wang YN, Song YH, Mišković ZL. Interactions of charged particle beams with double-layered two-dimensional quantum electron gases. Physics Letters A. 2014;**378**(22): 1626-1631. DOI: 10.1016/j.physleta.2014.04.007

[38] Borka D, Radović I, Vuković K. Energy loss of charged particles traversing multilayer graphene. Nuclear Instruments and Methods in Physics Research Section B: Beam Interactions with Materials and Atoms. 2015;**347**: 7-10. DOI: 10.1016/j.nimb.2015.01.057

[39] Song YH, Zhao D, Wang YN. Kinetic study on self-energy and stopping power of charged particles moving in metallic carbon nanotubes. Physical Review A. 2008;**78**:012901. DOI: 10.1103/PhysRevA.78.012901

[40] Zhao D, Song YH, Wang YN. Kinetic study on Channelling of protons in metallic carbon nanotubes. Chinese Physics Letters. 2008;**25**(7):2588. DOI: 10.1088/0256-307X/25/7/070

[41] Zhang YY, Sun JZ, Song YH, Mišković ZL, Wang YN. Channeling of protons in single-walled carbon nanotubes based on kinetic and molecular-dynamics treatment. Carbon. 2014;**71**:196-205. DOI: 10.1016/j.carbon.2014.01.030

[42] Fermanian Kammerer C, Méhats F. A kinetic model for the transport of electrons in a graphene layer. Journal of Computational Physics. 2016;**327**: 450-483. DOI: 10.1016/j.jcp.2016.09.010

[43] Zhang YY, Song YH, Radović I, Wang YN, Mišković ZL. Interactions of moving charged particles with triple-walled carbon nanotubes. European Physical Journal D: Atomic, Molecular, Optical and Plasma Physics. 2017;**71**:219. DOI: 10.1140/epjd/e2017-70744-3

[44] Loche P, Ayaz C, Schlaich A, Bonthuis DJ, Netz RR. Breakdown of linear dielectric theory for the interaction between hydrated ions and graphene. The Journal of Physical Chemistry Letters. 2018;**9**(22): 6463-6468. DOI: 10.1021/acs.jpclett.8b02473

[45] Hakimi S, Nguyen T, Farinella D, Lau CK, Wang HY, Taborek P, et al. Wakefield in solid state plasma with the ionic lattice force. Physics of Plasmas. 2018;**25**(2):023112. DOI: 10.1063/1.5016445

[46] Hakimi S, Zhang X, Lau C, Taborek P, Dollar F, Tajima T. X-ray laser Wakefield acceleration in a nanotube. International Journal of Modern Physics A. 2019;**34**(34): 1943011. DOI: 10.1142/S0217751X19430115

[47] Arista NR. Interaction of ions and molecules with surface modes in cylindrical channels in solids. Physical Review A. 2001;**64**:032901. DOI: 10.1103/PhysRevA.64.032901

[48] Martín-Luna P, Resta-López J. Excitation of plasmonic wakefields in multi-walled carbon nanotubes: A hydrodynamic approach. In: Kuznetsov DA, editor. Carbon Nanotubes - Recent Advances, Perspectives and Applications. Rijeka: IntechOpen; 2024. DOI: 10.5772/intechopen.114270

[49] Östling D, Tománek D, Rosén A. Electronic structure of single-wall, multiwall, and filled carbon nanotubes. Physical Review B. 1997;**55**:13980-13988. DOI: 10.1103/PhysRevB.55.13980

Section 2

# New Ideas for Electromagnetic Functional Materials and Devices

Chapter 4

# Homogenizing Electromagnetic Fields for Microperiodic Dielectrics

*Jolanta M. Wołowicz and Ryszard Wojnar*


## Abstract

Homogenization of the electromagnetism equations expressed by two fields, a scalar potential and a vector potential in a microperiodic inhomogeneous dielectric, is carried out. Both fields after homogenization are coupled, unlike in the case of a homogeneous medium, when they are separated. The homogenized medium is characterized by three coefficients: the effective dielectric coefficient, the average dielectric coefficient, and the root mean square dielectric coefficient. Two examples of determining the electrical permittivity are given.


**Keywords:** scalar and vector potentials, effective dielectric coefficient, cellular medium, asymptotic homogenization, chessboard composite

## 1. Introduction

Classical electrodynamics [1–3] in a dielectric with a cellular structure is considered. The length of the cell is small compared to the size of the medium and the length of the electromagnetic wave, and therefore we speak of a microperiodic medium. We use the method of homogenization of microperiodic media described in books [4–7]. We have used this method earlier to solve partial electrical problems related to piezoelectricity and magneto-electro-elastic media [8–10].

There are other methods for finding the effective constants of a non-homogeneous medium: the method of Γ-convergence [11] or based upon our compact group approach and the Hashin-Shtrikman variational theory proposed in [12], cf. also [13, 14].

The homogenization method describes fields in a non-homogeneous cellular medium using two independent spatial variables, a macroscopic variable $x$ and a microscopic variable $y$. The smallness parameter $\lambda$ allows us to isolate and describe the fields prevailing inside the cell. The fields also depend on the time variable $t$, but homogenization is not performed with respect to this variable.

In order to homogenize the electromagnetic field, we apply the formulation of the field equations using the potentials: vector $A$ and scalar $\varphi$. Unlike the original Maxwell equations, which are first-order partial equations, the potential equations are second-order partial equations, which facilitates their homogenization.

We used this type of approach in Ref. [15] to describe the behavior of an electromagnetic wave in a medium with a variable coefficient of dielectric permeability and magnetic susceptibility. However, in this case we were unable to prove that the zero



IntechOpen

term $A^{(0)}$ of the vector field expansion does not depend on the microscopic variable y; we accepted this fact ad hoc. In this work we limit ourselves to a dielectric medium for which we can assume that the magnetic susceptibility is constant, $\mu = 1$. With such a restriction we can prove that not only the zero term $\varphi^{(0)}$ of the scalar potential expansion but also the zero term $A^{(0)}$ of the vector potential expansion do not depend on y.

In a body placed in an electric field, dipole moments are created. In a dielectric, the shifting of charges under the influence of the field takes place only within the molecule. Each molecule under the influence of the field becomes a dipole. There are also known dielectrics whose particles are naturally dipoles. These are molecules with an ionic bond, e.g., a water molecule. In a dielectric of this type not placed in an electric field, the directions of the dipole moments are scattered chaotically. As a result, the total dipole moment of such a dielectric is zero. The electric field arranges the dipoles and aligns them in its direction. From this interpretation, a conclusion is drawn as to the meaning of the term "microperiodic." The size of the microperiod must be large enough not to reveal the individual properties of the molecules and their dipoles.

In the next section, we give equations for potentials in a non-homogeneous medium, and in Section 3 we perform homogenization of our problem. In Section 3.1, we give the principles of homogenization theory. In Section 3.2, we deal with the homogenization of the Lorentz condition, and in Sections 3.3 and 3.4, we deal with the homogenization of potential fields. Due to the coupling of equations, we carry out homogenization gradually, referring at each stage to the results from neighboring points. In Section 4, we collect the most important results and draw some conclusions. Finally, we provide two examples of determining the effective electrical permittivity.

Initially, we conduct considerations in general, taking into account the possibility of changing magnetic permeability, but in the next steps we limit ourselves to the dielectric.

## 2. Basic equations

Let us consider a non-homogeneous body with a cellular structure characterized by the electrical permittivity $\varepsilon$, the magnetic permeability $\mu$, and the electric conductivity $\sigma$, which vary from point to point according to the periodic structure of the material. The coefficients $\varepsilon, \mu$, and $\sigma$ are assumed to be positive. The position is given by vector $x = (x_1, x_2, x_3,)$ and the time by $t$.

The electrodynamics equations in differential form in the Gaussian system of units are as follows:

$$\epsilon_{kls}\, E_{s,l} = -\frac{1}{c}\frac{\partial B_k}{\partial t} \tag{1}$$

$$\epsilon_{kls}\, H_{s,l} = \frac{1}{c}\frac{\partial D_k}{\partial t} + \frac{4\pi}{c} j_k \tag{2}$$

$$B_{k,k} = 0 \tag{3}$$

$$D_{k,k} = 4\pi\rho \tag{4}$$

Here $E$ and $H$ are the electric and magnetic fields, $D$ and $B$ are the electric and magnetic inductions, while $\rho$ and $j$ denote the electric charge and current densities, respectively. The constant $c$ is the velocity of light in a vacuum.

The partial derivative of function $f = f(x, t)$ with respect to spatial coordinate is denoted by a comma $f_{,k} = \frac{\partial f}{\partial x_k}$ and the repeated indices mean summation $f_{,kk} = \sum_{k-1}^{3} \frac{\partial^2 f}{\partial x_k \partial x_k}$

The Levi-Civita completely antisymmetric symbol $\epsilon_{ijk}$ is plus 1, if $(i, j, k)$ is an even permutation of (1, 2, 3), minus 1 if it is an odd permutation, and 0 if any index is repeated. The following relation is important: $\epsilon_{abs}\ \epsilon_{spq} = \delta_{ap}\ \delta_{ap} - \delta_{aq}\ \delta_{bp}$

Eqs. (1)–(4) are supplemented by a description of the material properties of the medium:

$$D_s = \varepsilon \tag{5}$$

$$B_s = \mu\, H_s \tag{6}$$

$$j_s = \sigma\, E_s \tag{7}$$

In addition, two more potentials are introduced, the vector potential $A$ and the scalar potential $\varphi$ with the relations:

$$B_s = \epsilon_{slk}\, A_{k,l} \tag{8}$$

$$E_k = -\frac{1}{c}\frac{\partial A_k}{\partial t} - \varphi_{,k} \tag{9}$$

These potentials satisfy Eqs. (1) and (3) identically, and Eqs. (2) and (4) expressed in terms of the potentials take the form:

$$\left(\frac{1}{\mu}\, A_{k,l}\right)_{,l} - \left(\frac{1}{\mu}\, A_{l,k}\right)_{,l} - \frac{\varepsilon}{c}\left(\frac{1}{c}\frac{\partial^2 A_k}{\partial t^2} + \frac{\partial\, \varphi_{,k}}{\partial t}\right) = -\frac{4\pi}{c}\, j_k \tag{10}$$

$$\varepsilon_{,k}\left(\frac{1}{c}\frac{\partial A_k}{\partial t} + \varphi_{,k}\right) + \varepsilon\left(\frac{1}{c}\frac{\partial A_{k,k}}{\partial t} + \varphi_{,kk}\right) = -4\pi\rho \tag{11}$$

The current density vector $j$ can be considered either as given by relation (7) or as a given vector function of position $x$ and time $t$. The charge density $\rho$ should be considered as a given function of $x$ and $t$.

The potentials $A$ and $\varphi$ are not uniquely defined by formulas (8) and (9). It is seen from them that the divergence of vector $A$ can be assumed arbitrarily. Usually, the Lorentz condition is assumed:

$$A_{k,k} + \frac{\varepsilon\,\mu}{c}\frac{\partial\varphi}{\partial t} = 0 \tag{12}$$

If the potentials satisfy this condition, we can write Eqs. (10) and (11) in the form:

$$-\frac{\partial(1/\mu)}{\partial x_l}\frac{\partial A_l}{\partial x_k} + \frac{1}{\mu}\frac{\partial(\varepsilon\mu/c)}{\partial x_k}\frac{\partial\varphi}{\partial t} + \frac{\partial}{\partial x_l}\left(\frac{1}{\mu}\frac{\partial A_k}{\partial x_l}\right) - \frac{\varepsilon}{c^2}\frac{\partial^2 A_k}{\partial t^2} = -\frac{4\pi}{c}\, j_k \tag{13}$$

$$\frac{1}{c}\frac{\partial\varepsilon}{\partial x_k}\frac{\partial A_k}{\partial t} + \frac{\partial}{\partial x_k}\left(\varepsilon\frac{\partial\varphi}{\partial x_k}\right) - \frac{\varepsilon^2\,\mu}{c^2}\frac{\partial^2\varphi}{\partial t^2} = -4\pi\rho \tag{14}$$

This is a system of two equations for $A_k$ and $\varphi$. In the case of a homogeneous medium, the two equations separate into two inhomogeneous wave equations, one for the vector potential and the other for the scalar one, as shown in Eqs. (55) and (57).

We will homogenize these equations, i.e., we will give the material coefficients of a homogeneous medium approximating a microperiodic medium.

In our considerations, all scalar and vector fields depend on position $x$ and time $t$. We do not reveal the dependence on time $t$ in the notation to save space.

## 3. Homogenization

### 3.1 Description of the method

The microperiodic material is a bounded set $\overline{\Omega}$ of three-dimensional space (the bar denotes the closure of the set), where $\Omega$ is a sufficiently regular domain with the boundary $\partial\Omega$.

We assume that the medium reveals a microperiodic structure, cf. [15, 16].

In the sequel, we apply the method of two-scale asymptotic expansion [1–4].

Let a microperiodic structure of body considered be $\lambda Y$ -periodic, where $\lambda > 0$ is a small parameter and $Y = \prod_{i=1}^{3}(0, y_i)$ is the so-called basic cell. For a fixed $\lambda$ the material functions:

$$\varepsilon^{\lambda} = \varepsilon\left(\frac{x}{\lambda}\right), \quad \mu^{\lambda} = \mu\left(\frac{x}{\lambda}\right), \qquad \sigma^{\lambda} = \sigma\left(\frac{x}{\lambda}\right), \quad x \in \Omega \tag{15}$$

are $\lambda Y$–periodic.

We introduce a microscopic variable:

$$y = \frac{x}{\lambda} \tag{16}$$

Then material functions are: $\varepsilon^{\lambda} = \varepsilon(y), \ \mu^{\lambda} = \mu(y), \qquad \sigma^{\lambda} = \sigma(y), \ y \in Y$

Additionally, we observe that for a quasi-periodic structure we would have:

$$\varepsilon^{\lambda} = \varepsilon(x,y), \ \mu^{\lambda} = \mu(x,y), \qquad \sigma^{\lambda} = \sigma(x,y), \quad x \in \Omega, \quad y \in Y \tag{17}$$

where the functions $\varepsilon^{\lambda} = \varepsilon(x,.), \ \mu^{\lambda} = \mu(x,.), \qquad \sigma^{\lambda} = \sigma(x,.)$ are $y$-periodic, and $x \in \Omega$ is the macroscopic variable.

The basic cell Y has the form of a cube. Its volume is $|Y|$. The domain $\Omega$ is assumed to have an $\lambda Y$-periodic structure. The set $\Omega$ is covered with a regular mesh of size $\lambda$, each cell being a cube $Y$.

From the mathematical point of view, homogenization means the passage with $\lambda$ to zero in an appropriate meaning, cf. [4, 5].

According to the two-scale asymptotic approach, instead of one space variable $x$, we introduce two variables, macroscopic $x$ and microscopic $y$, where $y = x/\lambda$, and instead of a $f(x)$, consider the function $f(x, y)$. Taking into account the formula for the total derivative (known as the chain rule), we have:

$$\frac{\partial f(x,y)}{\partial x_i} = \frac{\partial f(x,y)}{\partial x_i} + \frac{1}{\lambda}\frac{\partial f(x,y)}{\partial y_i} \qquad \text{with} \ \ y = \frac{x}{\lambda} \tag{18}$$

Finally, the asymptotic expansion is introduced, e.g.:

$$f(x) = f^{(0)}(x,y) + \lambda f^{(1)}(x,y) + \lambda^2 \; f^{(2)}(x,y) + \ldots \tag{19}$$

where functions $f^{(i)}, i = 0, 1, 2, \ldots$ are Y-periodic.

According to the method of asymptotic expansions, we compare the terms associated with the same power of $\lambda$.

### 3.2 Homogenization of the Lorentz condition

Let us start the homogenization from the Lorentz condition (12). We now replace the spatial partial derivatives according to the chain rule (18) and expand the potential into series by powers of $\lambda$. We get

$$\begin{aligned}&\left(\frac{\partial}{\partial x_k}\right) + \frac{1}{\lambda}\frac{\partial}{\partial y_k}\Big)\left(A_k^{(0)}(x,y) + \lambda\, A_k^{(1)}(x,y) + \lambda^2\, A_k^{(2)}(x,y) + \ldots\right)\\ &\quad + \frac{\varepsilon(y)\mu(y)}{c}\left(\varphi^{(0)}(x,y) + \lambda\varphi^{(0)}(x,y) + \lambda^2\varphi^{(0)}(x,y) + \ldots\right) = 0\end{aligned} \tag{20}$$

We equate the terms of the order $1/\lambda$ and obtain

$$\frac{\partial A_k^{(0)}(x,y)}{\partial y_k} = 0 \tag{21}$$

We will use this result in the next section to prove that the zero component, $\varphi^{(0)}$ of the scalar potential expansion does not depend on $y$.

The terms of order $\lambda^0 = 1$ are

$$\frac{\partial\, A_k^{(0)}(x)}{\partial x_k} + \frac{\partial A_k^{(1)}(x,y)}{\partial y_k} + \frac{\varepsilon(y)\mu(y)}{c}\,\frac{\partial\varphi^{(0)}(x)}{\partial t} = 0 \tag{22}$$

In the last equation we used the results (27) and (40) about the independence of $\varphi^{(0)}$ and $A_k^{(0)}$ from the variable $y$. Therefore, this equation is valid for dielectrics.

### 3.3 Homogenization of the scalar potential equation

According to the homogenization rules presented above, we replace in Eq. (14) the spatial partial derivatives according to the chain rule (18) and expand the potential into series by powers of $\lambda$.

$$\begin{aligned}&\frac{1}{\lambda}\frac{1}{c}\frac{\partial\varepsilon(y)}{\partial y_k}\frac{\partial}{\partial t}\left(A_k^{(0)}(x,y) + \lambda A_k^{(1)}(x,y) + \lambda^2\, A_k^{(2)}(x,y) + \ldots\right)\\ &+\left(\frac{\partial}{\partial x_k} + \frac{1}{\lambda}\frac{\partial}{\partial y_k}\right)\left[\varepsilon(y)\left(\frac{\partial}{\partial x_k} + \frac{1}{\lambda}\frac{\partial}{\partial y_k}\right)\left(\varphi^{(0)}(x,y) + \lambda\varphi^{(1)}(x,y) + \lambda^2\varphi^{(2)}(x,y)\ldots\right)\right]\\ &- \frac{\varepsilon(y)^2\mu(y)}{c^2}\frac{\partial^2}{\partial t^2}\left(\varphi^{(0)}(x,y) + \lambda\,\varphi^{(1)}(x,y) + \lambda^2\,\varphi^{(2)}(x,y) + \;\ldots\;\right) = -4\pi\,\rho(y)\end{aligned} \tag{23}$$

According to the method of asymptotic expansions, we successively compare the terms associated with the same power of $\lambda$. We obtain at $\lambda^{-2}$

$$\frac{\partial}{\partial y_k}\left[\varepsilon(y)\,\frac{\partial\,\varphi^{(0)}(x,y)}{\partial y_k}\right]=0 \tag{24}$$

We multiply the last equation by $\varphi^{(0)}$ on both sides, integrate by parts over the Y cell, use periodic boundary conditions on the cell, and get

$$\int_Y \varepsilon(y)\,\frac{\partial\varphi^{(0)}(x,y)}{\partial y_k}\,\frac{\partial\varphi^{(0)}(x,y)}{\partial y_k}\,\mathrm{d}\,Y=0 \tag{25}$$

The left-hand side is always positive. It is clear that the function $\varphi^{(0)}$ cannot depend on $y$,

$$\frac{\partial\varphi^{(0)}(x,y)}{\partial y_k}=0 \tag{26}$$

We therefore write

$$\varphi^{(0)}=\varphi^{(0)}(x) \tag{27}$$

According to this result, the zero term $\varphi^{(0)}$ of the expansion with respect to the parameter $\lambda$ does not depend on the microscopic variable $y$.

Now, in this section we restrict the considerations to dielectrics, for which $\mu=1$. At $\lambda^{-1}$ we get

$$\frac{1}{c}\frac{\partial\varepsilon(y)}{\partial y_k}\frac{\partial A_k^{(0)}(x)}{\partial t}+\frac{\partial}{\partial y_k}\left[\varepsilon(y)\left(\frac{\partial\varphi^{(0)}(x)}{\partial x_k}+\frac{\partial\varphi^{(1)}(x,y)}{\partial y_k}\right)\right]=0 \tag{28}$$

In this equation the result (40) of the next section is used, that $A_k^{(0)}$ does not depend on $y$.

This equation will be satisfied if we assume:

$$\varphi^{(1)}(x,y)=\alpha_l(\mathrm{y})\left[\frac{1}{c}\frac{\partial A_l^{(0)}(x)}{\partial t}+\frac{\partial\varphi^{(0)}(x)}{\partial x_l}\right] \tag{29}$$

Here the vector function $\alpha_l(\mathrm{y})$ is Y-periodic solution to the following local equation:

$$\frac{\partial}{\partial y_k}\left[\varepsilon(y)\left(\delta_{kl}+\frac{\partial\alpha_l(\mathrm{y})}{\partial y_k}\right)\right]=0 \tag{30}$$

Finally, the coefficients at $\lambda^0=1$ give

$$\frac{1}{c}\frac{\partial\varepsilon(y)}{\partial y_k}\frac{\partial A_k^{(1)}(x,y)}{\partial t}+\frac{\partial}{\partial x_k}\left[\varepsilon(y)\left(\frac{\partial\varphi^{(0)}(x)}{\partial x_k}+\frac{\partial\varphi^{(1)}(x,y)}{\partial y_k}\right)\right]-\frac{\varepsilon(y)^2}{c^2}\frac{\partial^2\varphi^{(0)}(x)}{\partial t^2}=0 \tag{31}$$

and after using the solution (29):

$$\frac{1}{c}\frac{\partial\varepsilon(y)}{\partial y_k}\frac{\partial A_k^{(1)}(x,y)}{\partial t}+\frac{\partial}{\partial x_k}\left[\varepsilon(y)\left(\delta_{kl}+\frac{\partial\alpha_l(\mathrm{y})}{\partial y_k}\right)\frac{\partial\varphi^{(0)}(x)}{\partial x_l}+\varepsilon(y)\frac{\partial\alpha_l(\mathrm{y})}{\partial y_k}\frac{\partial A_l^{(0)}(x)}{c\,\partial t}\right]$$
$$-\frac{\varepsilon(y)^2}{c^2}\frac{\partial^2\varphi^{(0)}(x)}{\partial t^2}=-4\pi\,\rho(y) \tag{32}$$

We have an identity

$$\frac{\partial\varepsilon(y)}{\partial y_k}\frac{\partial A_k^{(1)}(x,y)}{\partial t}=\frac{\partial}{\partial y_k}\left[\varepsilon(y)\frac{\partial A_k^{(1)}(x,y)}{\partial t}\right]-\varepsilon(y)\frac{\partial^2 A_k^{(1)}(x,y)}{\partial y_k\partial t}$$

Integrating by parts, we have

$$\int_Y\frac{\partial\varepsilon(y)}{\partial y_k}\frac{\partial A_k^{(1)}(x,y)}{\partial t}\mathrm{d}y=-\int_Y\varepsilon(y)\frac{\partial^2 A_k^{(1)}(x,y)}{\partial y_k\,\partial t}\mathrm{d}y$$

and according to Eq. (22):

$$\frac{\partial A_k^{(1)}(x,y)}{\partial y_k}=-\frac{\partial A_k^{(0)}(x)}{\partial x_k}-\frac{\varepsilon(y)}{c}\frac{\partial\varphi^{(0)}(x)}{\partial t} \tag{33}$$

After integrating (32) and rearranging the terms, we get

$$\int_Y\varepsilon(y)\left(\delta_{kl}+\frac{\partial\alpha_l(\mathrm{y})}{\partial y_k}\right)\mathrm{d}y\left[\frac{\partial^2\varphi^{(0)}(x)}{\partial x_k\partial x_l}+\frac{\partial^2 A_l^{(0)}(x)}{\partial x_k\,\partial t}\right]+\int_Y\frac{\varepsilon(y)^2}{c^2}\mathrm{d}y\left[\frac{\partial^2\varphi^{(0)}(x)}{\partial t^2}-\frac{\partial^2\varphi^{(0)}(x)}{\partial t^2}\right]$$
$$=-4\pi\int_Y\rho(y)\mathrm{d}y \tag{34}$$

### 3.4 Homogenization of the vector potential equation

We replace in Eq. (13) the spatial partial derivatives according to the chain rule (18) and expand the vector potential $A$ into a series by powers of $\lambda$.

$$-\frac{1}{\lambda}\frac{\partial[1/\mu(y)]}{\partial y_l}\left(\frac{\partial}{\partial x_k}+\frac{1}{\lambda}\frac{\partial}{\partial y_k}\right)\left(A_l^{(0)}(x,y)+\lambda A_l^{(1)}(x,y)+\lambda^2A_l^{(2)}(x,y)+\dots\right)$$
$$+\frac{1}{\lambda}\frac{1}{c\mu(y)}\frac{\partial[\varepsilon(y)\mu(y)]}{\partial y_k}\frac{\partial}{\partial t}\left(\varphi^{(0)}(x)+\lambda\varphi^{(1)}(x,y)+\lambda^2\,\varphi^{(2)}(x,y)+\dots\right)$$
$$+\left(\frac{\partial}{\partial x_l}+\frac{1}{\lambda}\frac{\partial}{\partial y_l}\right)\left[\frac{1}{\mu(y)}\left(\frac{\partial}{\partial x_l}+\frac{1}{\lambda}\frac{\partial}{\partial y_l}\right)\left(A_k^{(0)}+\lambda A_k^{(1)}+\lambda^2A_k^{(2)}+\dots\right)\right]$$
$$-\frac{\varepsilon(y)}{c^2}\frac{\partial^2}{\partial t^2}\left(A_k^{(0)}+\lambda A_k^{(1)}+\lambda^2A_k^{(2)}+\dots\right)=-\frac{4\pi}{c} \tag{35}$$

We now restrict homogenizing to fields in dielectrics for which the coefficient $\mu$ can be set equal to 1. Then Eq. (35) reduces to

$$\frac{1}{\lambda}\frac{1}{c}\frac{\partial[\varepsilon(y)]}{\partial y_k}\frac{\partial}{\partial t}\left(\varphi^{(0)}(x)+\lambda\varphi^{(1)}(x,y)+\lambda^2\varphi^{(2)}(x,y)+\ldots\right)$$
$$+\left(\frac{\partial}{\partial x_l}+\frac{1}{\lambda}\frac{\partial}{\partial y_l}\right)\left[\left(\frac{\partial}{\partial x_l}+\frac{1}{\lambda}\frac{\partial}{\partial y_l}\right)\left(A_k^{(0)}+\lambda A_k^{(1)}+\lambda^2A_k^{(2)}+\ldots\right)\right] \tag{36}$$
$$-\frac{\varepsilon(y)}{c^2}\frac{\partial^2}{\partial t^2}\left(A_k^{(0)}+\lambda A_k^{(1)}+\lambda^2A_k^{(2)}+\ldots\right)=-\frac{4\pi}{c}j_k$$

Again, compare the terms associated with the same power of $\lambda$. We obtain at $\lambda^{-2}$:

$$\frac{\partial^2 A_k^{(0)}(x,y)}{\partial y_l\partial y_l}=0 \tag{37}$$

Similarly, as before with the scalar potential, we multiply the last equation by $A_k^{(0)}$ on both sides, integrate by parts over the Y cell, use periodic boundary conditions on the cell, and get

$$\int_Y\frac{\partial A_k^{(0)}(x,y)}{\partial y_l}\frac{\partial A_k^{(0)}(x,y)}{\partial y_l}\,\mathrm{d}\,Y=0 \tag{38}$$

Under the integral it is the sum of non-negative components, and so it could be equal to zero; each component should vanish. Hence,

$$\frac{\partial A_k^{(0)}(x,y)}{\partial y_l}=0 \tag{39}$$

So $A_k^{(0)}$ does not depend on $y$,

$$A_k^{(0)}=A_k^{(0)}(x) \tag{40}$$

Next, taking into account the result (40), we equate the coefficients at $\lambda^{-1}$, and get

$$\frac{1}{c}\frac{\partial[\varepsilon(y)]}{\partial y_k}\frac{\partial\varphi^{(0)}(x)}{\partial t}+\frac{\partial}{\partial y_l}\left(\frac{\partial A_l^{(0)}(x)}{\partial x_k}+\frac{\partial A_l^{(1)}(x,y)}{\partial y_k}\right)=0 \tag{41}$$

or, evidently

$$\frac{1}{c}\frac{\partial[\varepsilon(y)]}{\partial y_k}\frac{\partial\varphi^{(0)}(x)}{\partial t}+\frac{\partial^2 A_l^{(1)}(x,y)}{\partial y_l\partial y_k}=0 \tag{42}$$

or

$$\frac{\partial}{\partial y_k}\left[\frac{\varepsilon(y)}{c}\frac{\partial\varphi^{(0)}(x)}{\partial t}+\frac{\partial A_l^{(1)}(x,y)}{\partial y_l}\right]=0 \tag{43}$$

But, from the Lorentz condition (22) for dielectrics,

$$\frac{\partial A_k^{(0)}(x)}{\partial x_k} + \frac{\partial A_k^{(1)}(x,y)}{\partial y_k} + \frac{\varepsilon(y)}{c}\frac{\partial \varphi^{(0)}(x)}{\partial t} = 0 \tag{44}$$

it is clear that the interior of the square bracket in Eq. (43) does not depend on $y$, and this equation is satisfied identically.

The coefficients at $\lambda^0 = 1$ give

$$\frac{1}{c}\frac{\partial \varepsilon(y)}{\partial y_k}\frac{\partial \varphi^{(1)}(x)}{\partial t} + \frac{\partial}{\partial x_l}\left(\frac{\partial A_k^{(0)}}{\partial x_l} + \frac{\partial A_k^{(1)}}{\partial y_l}\right) - \frac{\varepsilon(y)}{c^2}\frac{\partial^2 A_k^{(0)}(x)}{\partial t^2} = -\frac{4\pi}{c} j_k(x,y) \tag{45}$$

By the relation (29) we get

$$\begin{aligned}&\frac{1}{c}\frac{\partial \varepsilon(y)}{\partial y_k}\frac{\partial}{\partial t}\left\{\alpha_l(\mathrm{y})\left[\frac{1}{c}\frac{\partial A_l^{(0)}(x)}{\partial t} + \frac{\partial \varphi^{(0)}(x)}{\partial x_l}\right]\right\} + \frac{\partial}{\partial x_l}\left(\frac{\partial A_k^{(0)}}{\partial x_l} + \frac{\partial A_k^{(1)}}{\partial y_l}\right)\\ &-\frac{\varepsilon(y)}{c^2}\frac{\partial^2 A_k^{(0)}(x)}{\partial t^2} = -\frac{4\pi}{c} j_k(x,y)\end{aligned} \tag{46}$$

or

$$\begin{aligned}&\frac{1}{c}\frac{\partial \varepsilon(y)}{\partial y_k}\alpha_l(\mathrm{y})\frac{\partial^2 \varphi^{(0)}(x)}{\partial x_l \partial t} + \frac{\partial}{\partial x_l}\left(\frac{\partial A_k^{(0)}}{\partial x_l} + \frac{\partial A_k^{(1)}}{\partial y_l}\right)\\ &+\frac{1}{c^2}\left\{\frac{\partial}{\partial y_k}\left[\varepsilon(y)\alpha_l(\mathrm{y})\right] - \left[\varepsilon(y)\left(\delta_{kl} + \frac{\partial \alpha_l(\mathrm{y})}{\partial y_k}\right)\right]\right\}\frac{\partial^2 A_l^{(0)}(x)}{\partial t^2} = -\frac{4\pi}{c} j_k(x,y)\end{aligned} \tag{47}$$

Integrate both sides of the last equation over cell Y. After use of the periodic boundary,

$$\begin{aligned}&\frac{1}{c}\int_Y \frac{\partial \varepsilon(y)}{\partial y_k}\alpha_l(y)\,\mathrm{dy}\,\frac{\partial^2 \varphi^{(0)}(x)}{\partial x_l \partial t} + \frac{\partial^2 A_k^{(0)}}{\partial x_l \partial x_l} - \frac{1}{c^2}\int_Y\left[\varepsilon(y)\left(\delta_{kl} + \frac{\partial \alpha_l(\mathrm{y})}{\partial y_k}\right)\right]\mathrm{dy}\,\frac{\partial^2 A_l^{(0)}(x)}{\partial t^2}\\ &= -\frac{4\pi}{c}\int_Y j_k(x,y)\,\mathrm{dy}\end{aligned} \tag{48}$$

## 4. Results and conclusions

Let us summarize the most important results of our calculations. We introduce the abbreviation,

$$\langle(\ldots)\rangle = \frac{1}{|\mathrm{Y}|}\int_Y \mathrm{dy}\,(\ldots) \tag{49}$$

and notation for the effective dielectric coefficient,

$$\varepsilon_{kl}^{\mathrm{eff}} = \left\langle \varepsilon(y)\left(\delta_{kl} + \frac{\partial \alpha_l(\mathrm{y})}{\partial y_k}\right)\right\rangle \tag{50}$$

Now, the equation (34) is written in the form

$$\varepsilon_{kl}^{\text{eff}}\left[\frac{\partial^2\varphi^{(0)}(x)}{\partial x_k \partial x_l}+\frac{\partial^2 A_l^{(0)}(x)}{\partial x_k \partial t}\right]+\frac{\left\langle \varepsilon(y)^2\right\rangle}{c^2}\left[\frac{\partial^2\varphi^{(0)}(x)}{\partial t^2}-\frac{\partial^2\varphi^{(0)}(x)}{\partial t^2}\right]=-4\pi\left\langle \rho(y)\right\rangle \tag{51}$$

In essence of course, it is the equation

$$\varepsilon_{kl}^{\text{eff}}\left[\frac{\partial^2\varphi^{(0)}(x)}{\partial x_k \partial x_l}+\frac{\partial^2 A_l^{(0)}(x)}{\partial x_k \partial t}\right]=-4\pi\left\langle \rho(y)\right\rangle \tag{52}$$

We have kept the second square bracket in (51) for a moment to show the transition to the case of a homogeneous medium with dielectric constant $\varepsilon$.

Namely, in a homogeneous medium:

$$\left\langle \varepsilon(y)^2\right\rangle=\varepsilon^2 \tag{53}$$

and

$$\varepsilon_{kl}^{\text{eff}}=\varepsilon \tag{54}$$

Then,

$$\varepsilon_{kl}^{\text{eff}}\frac{\partial^2 A_l^{(0)}(x)}{\partial x_k \partial t}+\frac{\left\langle \varepsilon(y)^2\right\rangle}{c^2}\frac{\partial^2\varphi^{(0)}(x)}{\partial t^2}=\frac{\partial}{\partial t}\left[\varepsilon_{kl}^{\text{eff}}\frac{\partial A_l^{(0)}(x)}{\partial x_k}+\frac{\left\langle \varepsilon(y)^2\right\rangle}{c^2}\frac{\partial\varphi^{(0)}(x)}{\partial t}\right]=0$$

in virtue of Eqs. (53) and (54) and the Lorentz condition (12).

Eq. (51) then takes the form of the wave equation from the classical field theory for a homogeneous dielectric medium:

$$\frac{\varepsilon}{c^2}\frac{\partial^2\varphi}{\partial t^2}-\frac{\partial^2\varphi}{\partial x_k \partial x_k}=\frac{4\pi}{\varepsilon}\rho \tag{55}$$

We apply the same procedure to Eq. (48) describing vector potential and get:

$$\frac{1}{c}\left(\left\langle \varepsilon(y)\right\rangle\delta_{kl}-\varepsilon_{kl}^{\text{eff}}\right)\frac{\partial^2\varphi^{(0)}(x)}{\partial x_l \partial t}+\frac{\partial^2 A_k^{(0)}}{\partial x_l \partial x_l}-\frac{1}{c^2}\varepsilon_{kl}^{\text{eff}}\frac{\partial^2 A_l^{(0)}(x)}{\partial t^2}=-\frac{4\pi}{c}\left\langle j_k(x,y)\right\rangle \tag{56}$$

In a homogeneous medium, when Eqs. (53) and (54) hold true, we get the classical wave equation:

$$\frac{\varepsilon}{c^2}\frac{\partial^2 A_k}{\partial t^2}-\frac{\partial^2 A_k}{\partial x_l \partial x_l}=\frac{4\pi}{c}j_k \tag{57}$$

Eqs. (52) and (56) are our basic results. They show that in a non-homogeneous medium:

i. There is a coupling of the equations for the potentials $A$ and $\varphi$. Only in the homogeneous case are the equations for this field separated.

ii. The equations for both fields represent damped waves, which was, of course, to be expected.

iii. A microperiodically non-homogeneous medium is anisotropic, which is also obvious.

iv. Instead of one material constant, as in a homogeneous medium, we have three constants, one tensor $\varepsilon_{kl}^{\text{eff}}$ and two mean values $\langle\varepsilon(y)\rangle$ and $\left\langle\varepsilon(y)^2\right\rangle$.

## 5. Examples

### 5.1 One-dimensional case

As an example, let us first consider a layered medium consisting of two types of dielectrics, cf. **Figure 1**. The basic cell consists of two materials, and the problem is one-dimensional.

Eq. (30) for the local function now takes the form:

$$\frac{\mathrm{d}}{\mathrm{d}y}\left[\varepsilon(y)\left(1+\frac{\mathrm{d}\alpha(\mathrm{y})}{\mathrm{d}y}\right)\right]=0 \tag{58}$$

with

$$\varepsilon(y)=\begin{cases}\varepsilon_1 & \text{for} \quad 0<y<y_1\\ \varepsilon_2 & \text{for} \quad y_1\le y<Y\end{cases} \tag{59}$$

Integrating (58) with periodic boundary values gives:

$$\alpha(y)=\frac{\varepsilon_2-\varepsilon_1}{\varepsilon_1+\varepsilon_2}y \qquad \text{for} \quad 0<y<y_1 \tag{60}$$

$$\alpha(y)=\frac{\varepsilon_2-\varepsilon_1}{\varepsilon_1+\varepsilon_2}y_1+\frac{\varepsilon_1-\varepsilon_2}{\varepsilon_1+\varepsilon_2}y \qquad \text{for} \quad y_1\le y<Y \tag{61}$$

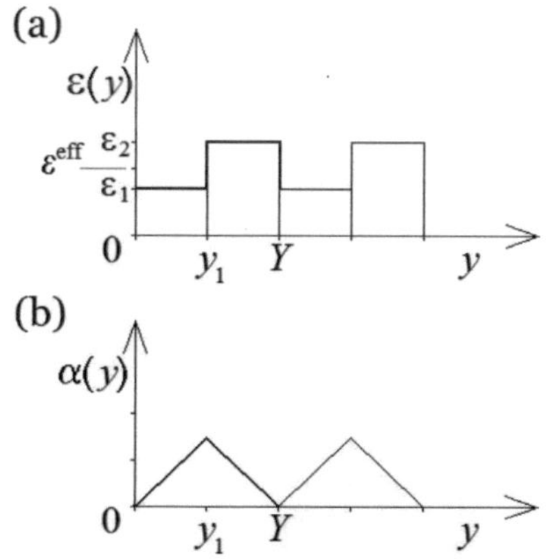


**Figure 1.**
*Layered medium of two dielectric layers with constants $\varepsilon_1$ and $\varepsilon_2$, (a). Local function $\alpha(\mathrm{y})$ for constants $\varepsilon_1=1$, $\varepsilon_2=2$, and $y_1=Y/2$ (b).*

The effective coefficient for our case after (50) reads:

$$\varepsilon^{\text{eff}} = \frac{1}{Y}\int_0^Y \text{d}y\ \varepsilon(y)\left(1+\frac{\text{d}\alpha(\text{y})}{\text{d}y}\right) \tag{62}$$

After performing the integration with $\alpha(y)$ given by (60) and (61) we get:

$$\varepsilon^{\text{eff}} = 2\ \frac{\varepsilon_1\varepsilon_2}{\varepsilon_1+\varepsilon_2} \tag{63}$$

The effective electrical permittivity is the harmonic mean of the values $\varepsilon_1$ and $\varepsilon_2$. This trivial example shows how the homogenization method works.

### 5.2 Two-dimensional case

Now let us consider a two-dimensional problem.

Let the dielectric coefficient have two values, $\varepsilon_1$ and $\varepsilon_2$ as before, but both values are distributed over the surface in a statistically equivalent way, i.e. the areas occupied by these values are the same. In particular, they can be distributed like the black and white squares on a chessboard, as shown in **Figure 2**. Let us write formula (30) in two dimensions

$$\frac{\partial}{\partial y_k}\left[\varepsilon(y)\left(\delta_{kl}+\frac{\partial\alpha_l(\text{y})}{\partial y_k}\right)\right] = 0 \tag{64}$$

Now the indices *k, l* take only two values, 1,2. Let us denote:

$$e_k^l = \delta_{kl} + \frac{\partial\alpha_l(\text{y})}{\partial y_k} \tag{65}$$

and

$$q_k^l = \varepsilon\ e_k^l \tag{66}$$

Based on (64) we have:

$$\frac{\partial\, q_k^l}{\partial y_k} = 0 \tag{67}$$

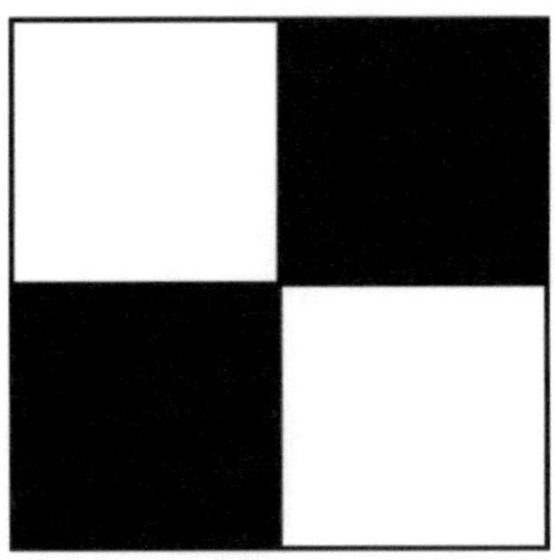

**Figure 2.**
*Basic cell of a chessboard composite, e.g., white square $\varepsilon_1$, black square $\varepsilon_2$.*

Similarly, using the curl operation:

$$\epsilon_{3ab} \frac{\partial e_b^l}{\partial y_a} = 0 \tag{68}$$

Because,

$$\epsilon_{3ab} \frac{\partial}{\partial y_a} \left( \delta_{bl} + \frac{\partial \alpha_l (\mathrm{y})}{\partial y_b} \right) = 0 \tag{69}$$

Eqs. (66), (67) and (68) correspond formally at the microscopic level to the macroscopic Eqs. (5), (4) and (1), respectively.

Eqs. (67) and (68) can be written as follows:

$$q_{1,1}^l + q_{2,2}^l = 0 \tag{70}$$

and

$$e_{2,1}^l - e_{1,2}^l = 0 \tag{71}$$

where the comma denotes partial differentiation with respect to the indicated component.

We are introducing new vectors:

$$q_a^{l*} = A\, R_{ab}\; e_b^l \tag{72}$$

and

$$e_a^{l*} = \frac{1}{A} R_{ab}\; q_b^l \tag{73}$$

where $A$ is an unknown constant, and

$$R_{ab} = \begin{vmatrix} 0 & -1 \\ 1 & 0 \end{vmatrix} \tag{74}$$

We verify

$$q_a^{l*} = A\, R_{ab}\; e_b^l = A \begin{vmatrix} 0 & -1 \\ 1 & 0 \end{vmatrix} \bullet \begin{vmatrix} e_1^l \\ e_2^l \end{vmatrix} = A\left(-e_2^l, e_1^l\right) \tag{75}$$

and by (71)

$$q_{a,a}^{l*} = A\left(-e_{2,1}^l + e_{1,2}^l\right) = 0 \tag{76}$$

On the other hand

$$e_a^{l*} = \frac{1}{A} R_{ab}\; q_b^l = \frac{1}{A} \begin{vmatrix} 0 & -1 \\ 1 & 0 \end{vmatrix} \bullet \begin{vmatrix} q_1^l \\ q_2^l \end{vmatrix} = \frac{1}{A}\left(-q_2^l, q_1^l\right) \tag{77}$$

and by (70):

$$\epsilon_{3ab}\, e^{l\,*}_{b,a} = \frac{1}{A}\left(q^{l}_{1,1} + q^{l}_{2,2}\right) = 0 \tag{78}$$

Moreover, by (66)

$$e^{l}_{a} = \frac{q^{l}_{a}}{\varepsilon} \tag{79}$$

and by (77)

$$q^{l\,*}_{a} = A\, R_{ab}\; e^{l}_{b} = A^{2}\, \frac{1}{\varepsilon}\, \frac{1}{A}\; R_{ab}\, q^{l}_{b} = A^{2}\, \frac{1}{\varepsilon}\, e^{l\,*}_{a} \tag{80}$$

We write this result as

$$q^{l\,*}_{a} = \varepsilon^{*}\, e^{l\,*}_{a} \tag{81}$$

We choose $A$ in such a way that

$$\varepsilon^{*} = A^{2}\, \frac{1}{\varepsilon} = \begin{cases} \varepsilon_2 & \text{if} \quad \varepsilon = \varepsilon_1 \\ \varepsilon_1 & \text{if} \quad \varepsilon = \varepsilon_2 \end{cases} \tag{82}$$

it is

$$A^{2} = \varepsilon_1\, \varepsilon_2 \tag{83}$$

The system of equations for quantities with asterisks (81) and (76, 78) is the same as for quantities without asterisks (66) and (70, 71).

Now, consider the mean values of fields:

$$\langle q^{l}_{a} \rangle = \frac{1}{Y} \int_{Y} q^{l}_{a}\, \mathrm{d}Y \quad \text{and} \quad \langle e^{l}_{a} \rangle = \frac{1}{Y} \int_{Y} e^{l}_{a}\, \mathrm{d}Y \tag{84}$$

The effective dielectric coefficient $\varepsilon^{\mathrm{eff}}$ is the coefficient between means

$$\langle q^{l}_{a} \rangle = \varepsilon^{\mathrm{eff}}\; \langle e^{l}_{a} \rangle \tag{85}$$

In similar manner

$$\langle q^{l\,*}_{a} \rangle = \varepsilon^{\mathrm{eff}}\; \langle e^{l\,*}_{a} \rangle \tag{86}$$

and by (80) and (77)

$$A\, R_{ab} \langle\, e^{l}_{b} \rangle = \varepsilon^{\mathrm{eff}}\; \frac{1}{A}\, R_{ab}\; \langle q^{l}_{b} \rangle \tag{87}$$

Hence,

$$A\, \langle\, e^{l}_{b} \rangle = \varepsilon^{\mathrm{eff}}\; \frac{1}{A}\, \langle q^{l}_{b} \rangle \tag{88}$$

Comparing last result with (85) we get

$$\varepsilon^{\text{eff}} = A \tag{89}$$

and by (83)

$$\varepsilon^{\text{eff}} = \sqrt{\varepsilon_1\, \varepsilon_2} \tag{90}$$

In this way, starting from the two-dimensional local function equation (64), we have recovered Dykhne's result [16]. This particular example shows the possibilities inherent in the homogenization method.

## Author details

Jolanta M. Wołowicz and Ryszard Wojnar*
Institute of Fundamental Technological Research PAS, Warszawa, Poland

*Address all correspondence to: rwojnar@ippt.pan.pl

IntechOpen

© 2025 The Author(s). Licensee IntechOpen. This chapter is distributed under the terms of the Creative Commons Attribution License (http://creativecommons.org/licenses/by/4.0), which permits unrestricted use, distribution, and reproduction in any medium, provided the original work is properly cited.

## References


[1] Landau LD, Lifshitz EM. Electrodynamics of Continuous Media, Course of Theoretical Physics, Vol. 8. 2nd English ed., translated from the Russian by Sykes JB, Reid WH. Oxford—New York—Beijing—Frankfurt—Sao Paulo—Sydney—Tokyo—Toronto: Pergamon Press; 1987. 429 p

[2] Landau LD, Lifshitz EM. The Classical Field Theory, Course of Theoretical Physics, Vol. 2. 3rd English ed., translated from the Russian by Hamermesh M. Oxford—New York—Toronto—Sydney—Braunschweig: Pergamon Press; 1971. 387 p

[3] Suffczyński M. Elektrodynamika, in Polish. Warszawa: PWN; 1964. 457 p

[4] Bensoussan A, Lions JL, Papanicolau G. Asymptotic Analysis for Periodic Structures. Amsterdam: North-Holland; 1978. 392 p

[5] Sanchez-Palencia E. Non-homogeneous Media and Vibration Theory. Berlin—Heidelberg—New York: Springer Verlag; 1980. 400 p

[6] Sanchez-Hubert J, Sanchez-Palencia ES. Introduction aux methods asymptotiques et à l'homogénéisation. Paris: Masson; 1992. 266 p

[7] Bakhvalov NS, Panasenko GP. Homogenisation: Averaging Processes in Periodic Media: Mathematical Problems in the Mechanics of Composite Materials. Dordrecht—Boston—London: Kluwer; 1989. 366 p

[8] Gałka A, Telega JJ, Wojnar R. Homogenization and thermopiezolectricity. Mechanics Research Communications. 1992;**19**: 315-324

[9] Gałka A, Telega JJ, Wojnar R. Some computational aspects of homogenization of thermopiezolectric composites. Computer Assisted Mechanics and Engineering Sciences. 1996;**3**:133-154

[10] Mityushev V, Nosov D, Wojnar R. Two-dimensional equations of magneto-electro-elasticity. In: Andrianov I, Gluzman S, Mityushev V, editors. Mechanics and Physics of Structured Media: Asymptotic and Integral Equations Methods of Leonid Filshtinsky. London, San Diego CA, Cambridge MA, Oxford: Academic Press; 2022

[11] Telega JJ. Piezoelectricity and homogenization. Application to biomechanics. In: Maugin GA, editor. Continuum Models and Discrete Systems. Vol. 2. Harlow, Essex: Longman; 1991. p. 220

[12] Sushko M. Ya: Effective dielectric response of dispersions of graded particles. Physical Review E. 2017;**96**(6): 062121-062128

[13] Roach GF, Stratis IG, Yannacopoulos AN. Mathematical Analysis of Deterministic and Stochastic Problems in Complex Media Electromagnetics, Princeton Series in Applied Mathematics. Princeton and Oxford: Princeton University Press; 2012

[14] Mackay TG, Lakhtakia A. Modern Analytical Electromagnetic homogenization. San Rafael, CA: Morgan and Claypool Publishers; 2015. DOI: 10.1088/978-1-6270-5427-0

[15] Wojnar R. On electromagnetic wave equations for a nonhomegenous microperiodic medium. In: Kähler U,

Reissig M, Sabadini I, Vindas J, editors. Analysis, Applications, and Computations, Research Perspectives, Proceedings of the 13th ISAAC Congress (International Society for Analysis, its Applications and Computation), Ghent, Belgium, 2021 (Trends in Mathematics). 1st ed. Cham: Birkhäuser Springer Nature Switzerland AG; 2023

[16] Dykhne AM. Conductivity of a two-dimensional two-phase system. Soviet Physics JETP. 1971;**32**(1):63-65

Chapter 5

# An Electromagnetic Approach to Microstrip Structures

*Somdotta Roy Choudhury*


## Abstract

Microstrip transmission lines are widely used in RF (radio frequency) and microwave circuits, where the geometry and characteristics of the electric fields play an important role in the performance of the system. The electric field distribution in a microstrip structure depends on several factors, including the dimensions of the strip, the substrate material, and the frequency of operation. The electric fields in a microstrip transmission line are primarily confined between the conductor (the microstrip strip) and the ground plane beneath the dielectric substrate. The distribution of the electric field depends on whether the microstrip is considered as a quasi-TEM (transverse electromagnetic) line or a quasi-TM (transverse magnetic) line, which is more typical for higher frequencies. In this chapter, a new investigation is presented to exhibit EM characteristics of a special type of microstrip structure, which is called defected microstrip structure (DMS). The DMS structures have natural tendencies to provide filtering characteristics. By changing the geometrical shape of DMS structures, different kinds of novel band-reject filters are designed. First, a rectangular complementary split ring-shaped DMS filter is developed to provide a single narrowband bandstop response at the ISM band. Different parametric studies, circuit model representations, and surface current distributions are shown here. Another dual-band bandstop filter is designed to consist of a folded rectangular complementary split ring structure. This compact, tunable filter is suitable for ISM bands, WLAN bands, and WiMax bands applications.


**Keywords:** EM characteristics of microstrip, surface current, L-C equivalent circuits of microstrip, defected microstrip structures, Bandstop filter

## 1. Introduction

In standard microstrip lines, electromagnetic waves are confined within the dielectric substrate and guided along the conducting trace. When defects are introduced into the structure, the waveguiding properties are altered and are reported as defected microstrip structures (DMS) [1–4]. The DMS can cause scattering, resonance, and wave confinement, affecting how signals propagate. The DMS is often used in the design of compact RF (radio frequency) and microwave components. By introducing a defect, the size of a circuit can be reduced while maintaining performance. The frequency response of the microstrip line can be tuned by varying the



IntechOpen

size and shape of the defect, allowing for specific frequency bands to be suppressed or passed through. The defect in the microstrip line can change the impedance profile of the transmission line. These defected structures can create localized electromagnetic fields. These can act as a resonator, enabling DMS to serve as bandstop or bandpass filters. DMS filter was first reported by Schiffman and Matthaei [1] in 1964 for strip line technology and referred to as a spur line filter. DMS is generally applied to exhibit bandstop responses. A DMS is another technique for designing bandstop filters. It is made by etching a uniform or non-uniform slit or pattern over the signal line. In a DMS structure, there is no etching in the ground plane. Like the DGS structures, the DMS unit is a non-periodic EBG structure. DMS is the modification of the spur line structure. This spur line structure was first introduced by Schiffman and Matthaei [1]. Bates adapted this technique in the microstrip medium by assuming the same phase velocities for even and odd modes [2]. Nguyen et al. [3] presented an equivalent circuit of the same section taking into account the different even and odd mode phase velocities. Similar to the DGS structure, DMS increases the electrical length of the microstrip line, and disturbs its current distribution, and the effective capacitance and inductance of a microstrip line increase [4–8]. If compared, a DGS occupies more area than a DMS due to a bigger etched pattern over the ground plane. From the response of the defected microstrip unit cell, it can be assured that it behaves like an L-C resonant circuit but shows a bigger quality factor than DGS [9–11]. A DMS [12–22] is more easily integrated with other microwave circuits and has an effectively reduced circuit size compared with DGSs. Simultaneously, a DMS exhibits the properties of slow-wave, rejecting microwaves in certain frequencies, and has an increasing electric length of certain circuits, which are similar to the well-known DGS but without any modification of the ground plane [23].

Some attractive features of DMS structure:

- Deeper rejection level
- Sharper roll off
- Compact circuit size
- Moderate rejection bandwidth (about 10%)

Electrical characteristics of DMS structure:

- Radiates significantly less than conventional shunt-stub and coupled-line filters.
- Virtually nondispersive, that is, the response repeats at almost exact odd multiples of the resonant frequency (f0).
- Decreases phase velocity and provides slow wave characteristics.
- Increases effective capacitance and inductance of the transmission line where it is placed.
- Provides bandstop characteristics.

Advantages of using DMS circuits:

- Structure is very much simple.
- Structure is very much compact.
- Provides deeper stopband characteristics.
- Having extremely small element values.

Applications of DMS structures:

- Planar resonator.
- High characteristic impedance transmission lines.
- Microwave circuits such as filter, coupler, divider/combiner, oscillator, power amplifier, and antenna performance improvements.
- Notch band in ultra-wideband antennas and bandpass filters.

## 2. Design of narrowband Bandstop filter using rectangular complementary split ring resonator (RCSR)

Bandstop filters are very much useful in many applications to suppress the higher harmonics and spurious signals. Spur line filters provide the basic building concept for the defected microstrip structure (DMS). Researchers have proposed different types of band reject filtering circuits using different geometrical-shaped DMS structures in recent years.

### 2.1 Electromagnetic analysis of RCSR-shaped DMS structure

**Figure 1** shows the configuration of the investigated unit cell, RCSR-shaped DMS, on the signal plane of the microstrip line. This type of structure is etched from the 50-ohm microstrip line of width, w = 6.7036 mm, considering the dielectric constant of the material is to be 2.5 and the substrate thickness 2.36 mm. The different dimensions of the unit RCSR are: length, a = 14.6 mm and width, b = 5.2 mm, split gap, g = 0.5 mm and slit width d = 0.5 mm. The simulated result of the structure using IE3D simulation software shows a narrowband one-pole bandstop response

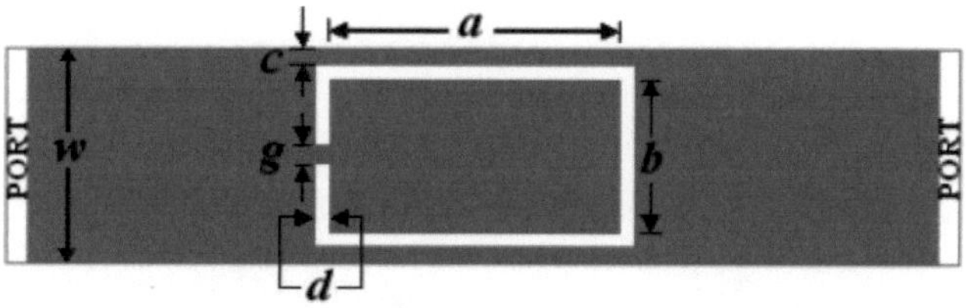


**Figure 1.**
*Schematic of proposed RCSRR structure.*

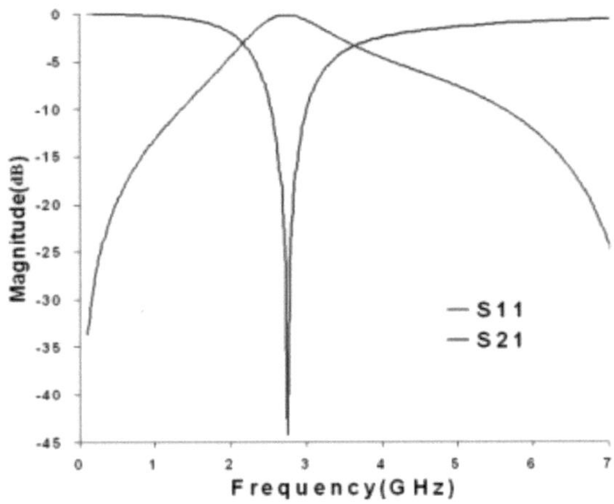


**Figure 2.**
*Simulated S-parameter response of the unit cell RCSRR.*

shown in **Figure 2**. The 3 dB lower and upper cutoff frequencies are fc1 = 2.1 GHz and fc2 = 3.7 GHz, respectively. The pole frequency (fp) is obtained at 2.7 GHz with a maximum rejection level of −45.5 dB. It is observed that the attenuation zero at 2.7GHz is very close to the attenuation pole location at 2.1 GHz. As a result, a high sharpness factor of 70.8 dB/GHz is obtained in the transition band. The −15 dB rejection bandwidth of the filter is 271 MHz.

## 2.2 Transmission line model of RCSR unit

**Figure 3** shows the equivalent circuit model of this RCSR-shaped DMS unit. This is obtained by dividing the line into four parts: two transmission lines (through lines) at the top and bottom of the complementary split ring having characteristic impedance Z1 and the electrical length θ1 each; the resonating stub line inside the rectangular split ring having characteristic impedance Z3 and electrical length θ3 and a transmission line corresponding to split gap "g1" having characteristic impedance Z2 and electrical length θ2. The capacitance "Cd" is introduced to describe the coupling capacitance between through line (Z1) and stub (Z3) [21].

## 2.3 Mathematical analysis of single unit

Using the network analysis theory, the s-parameters, S11 and S21 of the equivalent circuit model in **Figure 4** can be obtained as [21],

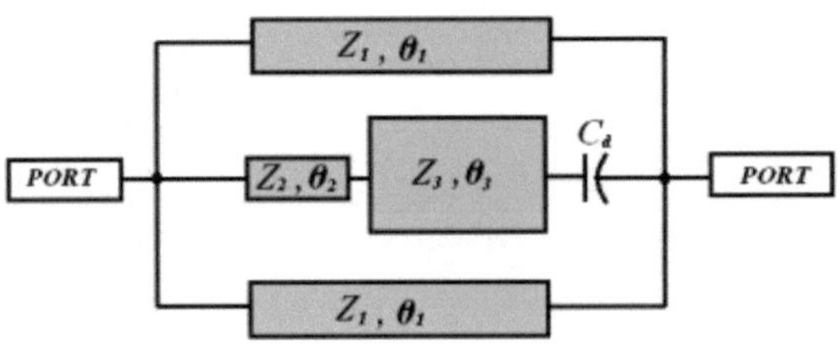


**Figure 3.**
*Equivalent circuit model of the unit cell RCSRR.*

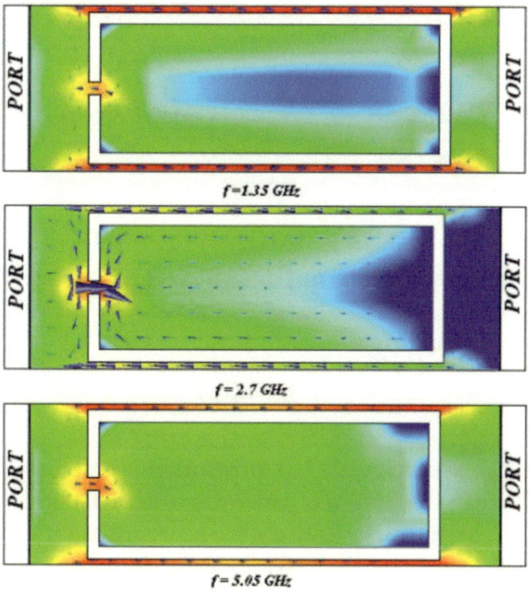


**Figure 4.**
*Simulated surface current distribution of investigating RCSR for different frequencies, (a) at below resonant frequency (1.35GHz), (b) at resonant frequency (2.7GHz), and (c) above resonant frequency (5.05GHz).*

$$S_{11} = \frac{(Y_0 - Y_{11})(Y_0 + Y_{22}) + Y_{12}Y_{21}}{(Y_0 + Y_{11})(Y_0 + Y_{22}) - Y_{12}Y_{21}} \tag{1}$$

$$S_{21} = \frac{-2Y_{12}Y_0}{(Y_0 + Y_{11})(Y_0 + Y_{22}) + Y_{12}Y_{21}} \tag{2}$$

$$Y_{11} = D_1 / B_1 - 2jY_1 \cot\theta_1 \tag{3}$$

$$Y_{12} = Y_{21} = -1 / B_1 + 2jY_1 \csc\theta_1 \tag{4}$$

$$Y_{22} = A_1 / B_1 - 2jY_1 \cot\theta_1 \tag{5}$$

where $Y_{11}$, $Y_{12}$, $Y_{21}$, and $Y_{22}$ are the admittance parameters and Y0 is the characteristic admittance of 50 ohm line, Y1 = 1/Z1. The ABCD parameters are:

$$A_1 = \cos\theta_2 \cos\theta_3 - Z_2Y_3 \sin\theta_2 \sin\theta_3 \tag{6}$$

$$B_1 = j(Z_2 \sin\theta_2 \cos\theta_3 + Z_3 \cos\theta_2 \sin\theta_3) + \frac{1}{j\omega C_d}(\cos\theta_2 \cos\theta_3 - Z_2Y_3 \sin\theta_2 \sin\theta_3) \tag{7}$$

$$C_1 = j\left(Y_2 \sin\theta_2 \cos\theta_3 + Y_3 \cos\theta_2 \sin\theta_3\right) \tag{8}$$

$$D_1 = \cos\theta_2 \cos\theta_3 - Y_2 Z_3 \sin\theta_2 \sin\theta_3 + \frac{1}{\omega C_d}\left(Y_2 \sin\theta_2 \cos\theta_3 + Y_3 \cos\theta_2 \sin\theta_3\right) \tag{9}$$

## 3. Design result of narrowband Bandstop filter using rectangular complementary split ring resonator (RCSR)

### 3.1 Surface current distribution

The surface current distribution of the RCSRR at below resonant frequency (1.35GHz), resonant frequency (2.7GHz) and above resonant frequency (5 GHz) are shown in **Figure 5**. It is observed that at resonance, the maximum surface current is confined within the gap "g1." At the frequencies of 1.35GHz and 5 GHz the maximum surface current distribution is observed in through lines (Z1, θ1), as evident from **Figure 4**.

### 3.2 Parametric study of the single unit

For a better understanding of the operation of the RCSRR, the effects of the dimensions on the transmission response are studied based on the following simulation results. It is observed that the stub length and width (*a* and b) of the investigating structure are directly dependent on the cutoff and pole frequencies (fc and fp). **Figures 5** and **6** show the influence of stub length "*a*" on the cutoff and pole frequencies. It is noticed that if "*a*" increases, the fc and fp decrease and vice versa.

Again, the variation of attenuation pole and attenuation zero with split gap "g1" for a fixed value of slot width "d" is depicted in **Figures 7** and **8**. It is seen from the figures that when "g1" increases, the cutoff and pole frequencies decrease and vice versa. It gives the relation between "g1" and the cutoff frequency and resonant frequency, respectively.

From the above discussion, it can be concluded that if either the stub length or width is increased, the perimeter of the RCSRR increases; simultaneously, the cutoff

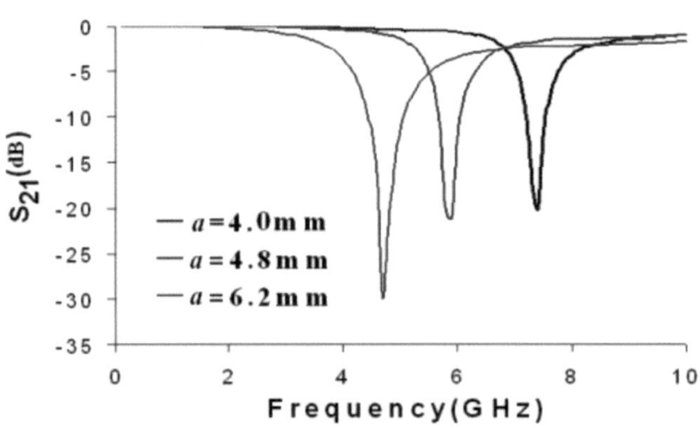


**Figure 5.**
*Variation of responses of rectangular split ring bandstop filter with different arm length "a."*

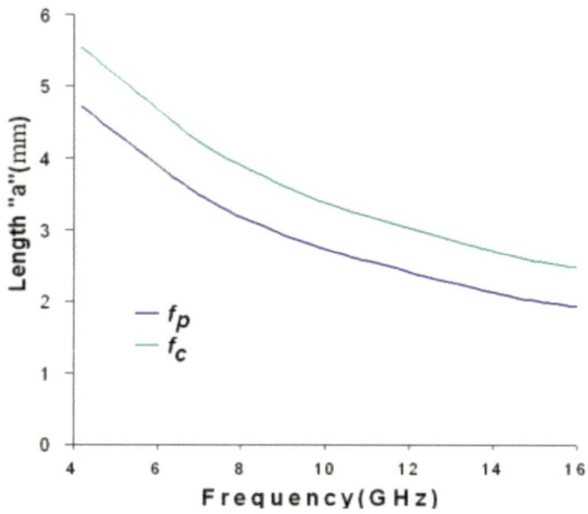


**Figure 6.**
*Variation of arm length "a" with cutoff frequency and pole frequency.*

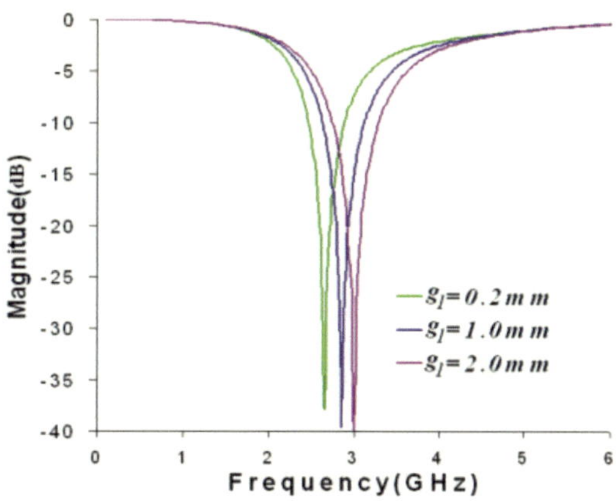


**Figure 7.**
*Variation of responses of rectangular split ring bandstop filter with different split gap.*

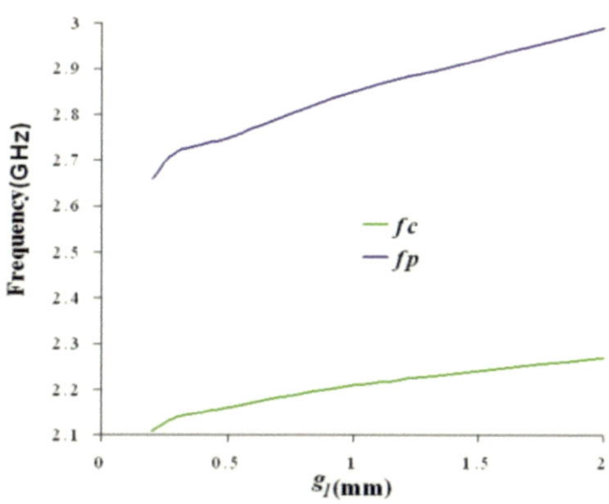


**Figure 8.**
*Variation of split gap with cutoff frequency and pole frequency.*

frequency and pole frequency decrease, and vice versa. Therefore, instead of changing the total dimensions of the RCSRR, the tuning of frequency can be obtained by simply changing either "*a*" or "b."

Moreover, attenuation poles (fc) and attenuation zeros (fp) can be tuned with the variation of the value of the slot gap "d," keeping the other dimensions fixed. **Figures 9** and **10** show the variation of cutoff and pole frequencies with different values of slot width "d" of the proposed structure. From **Figures 10** and **11**, it is possible to find the cutoff and pole frequencies for the specified value of the slot width, "d."

**Figure 11** describes the variation of gap capacitance "Cd" with the change of widths "b," "c," and the slot gap "d." It is observed that for a particular substrate, "Cd" increases as "d" decreases and "b" increases.

### 3.3 Equivalent circuit model of the RCSR unit

From the structure of RCSRR, it is seen that there is a capacitive effect due to the slot gap and an inductive effect due to current rotation around the rectangular-shaped slotted section. Therefore, the equivalent LC circuit of the proposed structure is given in **Figure 12**, and the comparison between EM-simulated and circuit-simulated responses

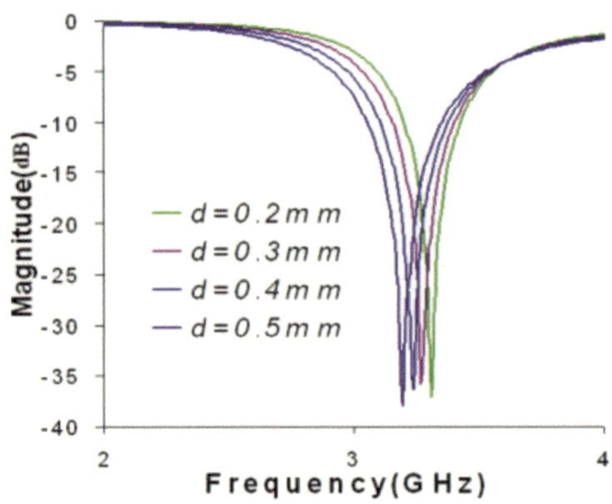


**Figure 9.**
*Variation of responses of rectangular split ring bandstop filter with different slot width.*

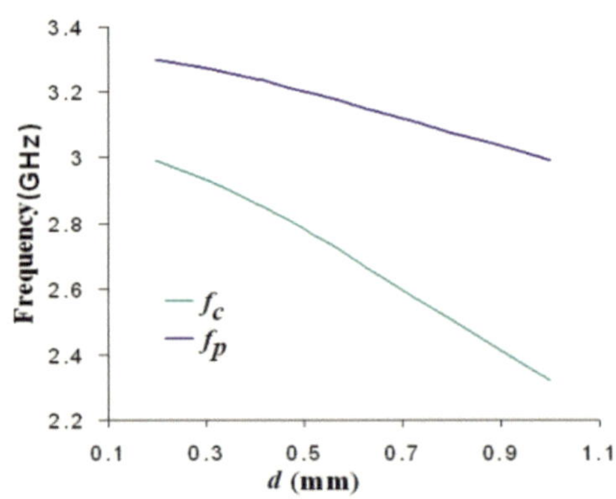


**Figure 10.**
*Variation of slot width with cutoff frequency and pole frequency.*

*DOI: http://dx.doi.org/10.5772/intechopen.1010148*

is presented in **Figure 13**. The values of L and C are related by Eqs. (10) and (11) [21] and are obtained as C = 7.57 pF and L = 0.465nH for the cutoff (fc) and pole frequencies (fp) of 2.1GHz and 2.68GHz, respectively:

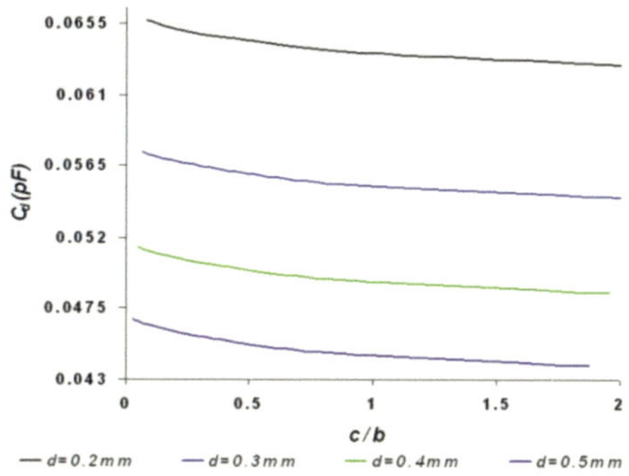


**Figure 11.**
*Gap capacitance "Cd" as functions of the gap "d" and the ratio of widths "b" and "c."*

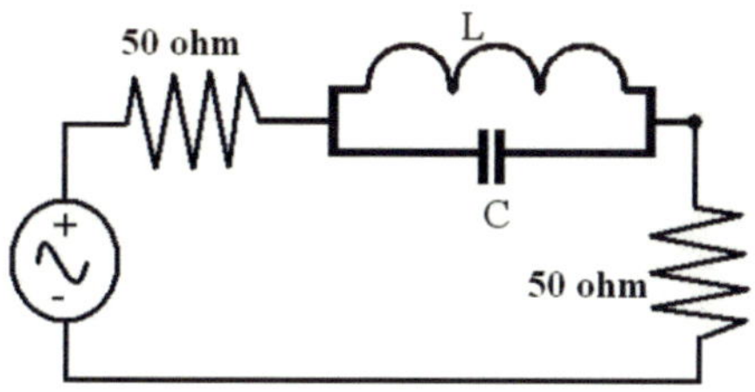


**Figure 12.**
*The equivalent LC circuit of the proposed RCSR structure.*

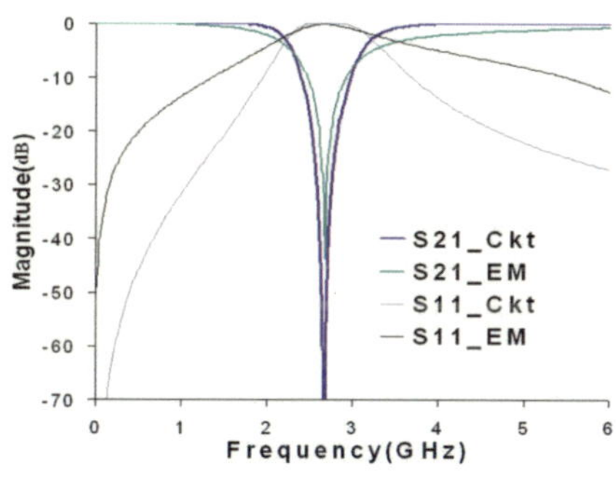


**Figure 13.**
*Comparison of EM-simulated and circuit-simulated responses for C = 7.57 pF and L = 0.465nH.*

$$C = \frac{f_c}{2Z_0\left(f_p^2 - f_c^2\right)} \tag{10}$$

$$L = \frac{1}{4\pi^2 f_p^2 C} \tag{11}$$

| References | Compactness ($\lambda_0 \times \lambda_0$) | Sharpness factor (dB/GHz) | Insertion loss at passband (dB) | Nature of stop bands |
|---|---|---|---|---|
| [24] | 0.024 × 0.28 | 270 | 2 | Single, narrow |
| [25] | 0.17 × 0.007 | 1681 | 1.2 | Single, narrow |
| [12] | 0.08 × 0.04 | 270 | 1.4 | Single, narrow |
| [13] | 0.13 × 0.10 | 4.7 | 1 | Single narrow |
| [14] | 0.23 × 0.02 | 74 | 1 | Single, narrow |
| Proposed Rectangular Complementary Split Ring (RCSR) shape | 0.13 × 0.05 | 70.8 | 0.03 | Single, narrow |

**Table 1.**
*Comparison of the investigated Filters with some literature available.*

### 3.4 Discussion on result of RCSR

**Table 1** provides some advantages of the proposed DMS structures compared to the state of the art. Due to the non-availability of the insertion loss data in some literature, the corresponding boxes remain blank.

### 3.5 Conclusions

The defected microstrip structures offer a flexible and efficient way to modify the electromagnetic properties of transmission lines, which can be used in various applications, from miniaturized circuits to high-performance RF and microwave components. The intentional introduction of defects provides a way to tailor electromagnetic behavior for specific needs, such as filtering, sensing, or impedance matching. The design and analysis of defected microstrip structures require the use of electromagnetic simulation tools. These tools, such as finite-difference time domain (FDTD), finite element method (FEM), and method of moments (MoM), can model the effect of defects on the transmission line. Here FEM method-enabled HFSS simulation has been done for the various results. In this chapter, a bandstop filter (narrowband) structure is designed using DMS. The novel RCSR-shaped DMS provides a single-stop-band response with a very compact structure size of 0.13 λ0 × 0.05 λ0.

*DOI: http://dx.doi.org/10.5772/intechopen.1010148*

## Author details

Somdotta Roy Choudhury
Ramra Adik Institute of Technology, Navi Mumbai, Maharashtra, India

*Address all correspondence to: somdottaroychoudhury@gmail.com

IntechOpen

© 2025 The Author(s). Licensee IntechOpen. This chapter is distributed under the terms of the Creative Commons Attribution License (http://creativecommons.org/licenses/by/4.0), which permits unrestricted use, distribution, and reproduction in any medium, provided the original work is properly cited.

## References

[1] Schiffman B, Matthaei G. Exact design of Bandstop microwave filters. IEEE Transactions on Microwave Theory and Techniques. 1964;**12**:6-15. DOI: 10.1109/TMTT.1964.1125744

[2] Bates R. Design of microstrip spur-line band-stop filters. IEE Journal on Microwaves, Optics and Acoustics. 1977;**1**:209-214. DOI: 10.1049/ij-moa.1977.0029

[3] Nguyen C, Chang K. On the analysis and design of spurline Bandstop filters. IEEE Transactions on Microwave Theory and Techniques. 1985;**33**:1416-1421. DOI: 10.1109/TMTT.19851133233

[4] Getsinger W. Dispersion of parallel-coupled microstrip. IEEE Transactions on Microwave Theory and Techniques. 1973;**21**:144-145. DOI: 10.1109/TMTT.1973.1127951

[5] Gould J, Talboys E. Even and odd-mode guide wavelengths of coupled lines in microstrip. Electronics Letters. 1972;**8**:121-122

[6] de Ronde F, Shammas S. MIC band filters using open-ring resonators. In: Proceedings of the 4th European Microwave Conference. Switzerland: IEEE MTT; Sept 1974. pp. 531-535

[7] Benedek P, Silvester P. Equivalent capacitances for microstrip gaps and steps. IEEE Transactions on Microwave Theory and Techniques. 1972;**20**:729-733

[8] Jose T-M, Jardon-Aguilar H. Comparison of defected ground structure (DGS) and defected microstrip structure (DMS) behavior at high frequencies. In: 1st International Conference on Electrical and Electronics Engineering. Acapulco, Mexico: IEEE; Sept 2004. DOI: 10.1109/ICEEE.2004.1433839

[9] Fallahzadeh S, Akbarzadeh A, Tayarani M. Spurious response suppression in microstrip bandpass filters using defected microstrip structures. Electromagnetics. 2012;**32**:389-400

[10] Wei F, Pei-Yuan Qin Y, Guo J, Shi X. Design of multiband bandpass filters based on stub loaded stepped-impedance resonator with defected microstrip structure. IET Microwaves, Antennas and Propagation. 2016;**10**:230-236. DOI: 10.1049/iet-map.2015.0495

[11] Chen L, Li X, Wei F. A compact quad-band bandpass filter based on defected microstrip structure. Frequenz. 2017;**71**:311-316. DOI: 10.1515/freq-2016-0238

[12] Xiao J, Zhu Y. Multi-band bandstop filter using inner T-shaped defected microstrip structure (DMS). International Journal of Electronics and Communications. 2014;**68**:90-96. DOI: 10.1016/j.aeue.2013.07.002

[13] Wang J, Ning H, Xiong Q, Mao L. A compact narrow-band Bandstop filter using spiral-shaped defected microstrip structure. Radioengineering. 2014;**23**:209-213

[14] Kazerooni M, Cheldavi A, Kamarei M. A novel bandpass defected microstrip structure (DMS) filter for planar circuits. In: Progress In Electromagnetics Research Symposium Proceedings. Russia; Aug 2015. pp. 1214-1217

[15] Zheng X, Pan Y, Jiang T. UWB bandpass filter with dual notched bands using T-shaped resonator

and L-shaped defected microstrip structure. Micromachines. 2018;**9**:1-11. DOI: 10.3390/mi9060280

[16] Boutejdar A, Challal M, Das S, Soumia E. Design and manufacturing of a novel compact 2.4 GHz LPF using a DGS-DMS combination and quasi octagonal resonators for radar and GPS applications. Progress In Electromagnetics Research C. 2019;**90**:15-28. DOI: 10.2528/PIERC18092107

[17] Mehraban S, Masoumi N, Safieddin S. A quasi-analytic synthesis methodology for spiral defected microstrip compact bandstop filters. International Journal of RF and Microwave Computer-Aided Engineering. 2020;**30**:1-11. DOI: 10.1002/mmce.22467

[18] Dokmetas B, Arican G-O, Akcam N, Yazgan E. A compact Bandstop filter design using DMS-DGS technique for radar applications. Applied Computational Electromagnetics Society Journal (ACES). 2021;**36**(11):1460-1467

[19] Shakeri S, Dousti M, Ghalamkari B. A quad-band defected microstrip structure loaded bandpass filter for standard GPS, WiFi, WLAN, and WiMAX applications. International Journal of RF and Microwave Computer-Aided Engineering. 2022;**32**:1-11. DOI: 10.1002/mmce.23082

[20] Weerathunge N, Chakraborty S, Mahon S. A compact power divider/combiner using a defected microstrip structure. In: Proceedings of the Australian Microwave Symposium; 16-17 February 2023. Melbourne, Australia: IEEE; 2023. pp. 31-32

[21] Miligy AF, Morsy ME, Elfeky AH, Fayad Y, Salama AA. Implementation of an enhanced antenna Array loaded with EMA/ADC and ITC-Egypt 2024 sign-shape slots defected microstrip structures (DMS). In: 2024 International Telecommunications Conference (ITC-Egypt); 22-25 July 2024. Cairo, Egypt: IEEE; 2024

[22] Gobinath A, Kumar NS, Rajeswari P, Anandan M. Hybrid curvilinear defected microstrip structure to reduce the electromagnetic coupling in PCB. In: 3rd International Conference for Innovation in Technology (INOCON); 01-03 March 2024. Bangalore, India: IEEE; 2024

[23] Hong J, Lancaster M. Microstrip Filters for RF/Microwave Applications. 2nd ed. New York: Willy; 2001

[24] Chongcheawchamnan M, Shafique M, Robertson I. Miniaturisation and electronic tuning techniques for microstrip spurline filters. IET Microwaves, Antennas and Propagation. 2011;**5**:1-9. DOI: 10.1049/iet-map.2010.0019

[25] Hari Krishna J. Compact spiral folded spurline for low frequency applications. Electronics Letters. 2011;**47**:1-2. DOI: 10.1049/el.2011.1508

Chapter 6

# Electromagnetic Materials: Shield, Meta-Surface Absorber, and Cloak

*Priyanka, Akshita Yadav, Aishwarya Pradeep, Himangshu B. Baskey and Prashant S. Alegaonkar*

## Abstract

In the advanced technological landscape, metamaterials have gained significant attention and demand owing to their astonishing properties that extend beyond what is typically observed in natural materials. These engineered materials possess the remarkable ability to manipulate electromagnetic waves in ways that conventional substances cannot. As a result, they have found numerous fascinating applications across various fields. For instance, metamaterials are utilized for electromagnetic interference shielding, offering advanced protection for electronic devices against disruptive signals. Additionally, they serve as highly efficient metasurface absorbers, capable of soaking up electromagnetic radiation with minimal reflection. Moreover, the innovative concept of cloaking technology leverages metamaterials to render objects virtually invisible, opening new frontiers in stealth applications. The interaction of electromagnetic radiation with matter has an interesting effect. Herein, we are discussing electromagnetic radiation, typically, in the range of microwave radar region. The interaction aspect has been discussed from the perspective of electromagnetic interference shielding, subwavelength metasurfaces, and invisible carpet cloaks. We have presented the data for a typical coating material, their inclusions, the architecture of the metasurfaces, and the design of cloaks. Their scattering properties, including relevant constitutive analysis, have been presented under limiting conditions.

**Keywords:** EMI material, metasurfaces, carpet cloak, constitutive analysis, stealth

## 1. Introduction

In wave optics, the phenomena of dispersion, interference, diffraction, and polarization date back to the early Newtonian era. Waves are generally classified into two primary categories: electromagnetic (EM) and mechanical waves. EM waves are unique in that they do not require a physical medium, like air or water, to propagate. This allows them to traverse empty space, transporting energy across vast distances without any obstacles. The capability to propagate without a medium unpins many technologies and applications, including satellite communications, radio astronomy, remote sensing, and many more [1]. When EM waves encounter a material medium, their interaction can lead to a variety of outcomes, including absorption, transmission, scattering, or reflection. Each of these interactions depends heavily on the specific

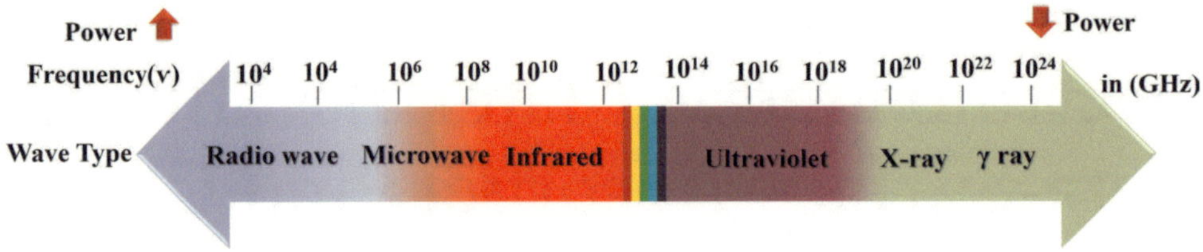


**Figure 1.**
*EM spectrum.*

characteristics of both the EM wave and the medium involved [2]. The EM wave spectrum can be broadly categorized as shown in **Figure 1**. As illustrated in **Figure 1**, this spectrum is divided into seven distinct segments. At the high-energy end of the spectrum are gamma rays, then X-rays, ultraviolet (UV), and so on [3].

### 1.1 Metamaterials - The novel approach

At the macroscopic level, the complex behaviors of individual atoms in a material play a crucial role in its EM properties. To understand how these atoms align and bond to form the material, we need to consider how thermodynamic laws influence their interactions and arrangements. When analyzing how materials interact with EM radiation, it is essential to focus on the averaged responses of the material over time and space. This is particularly important when the wavelength of a wave is significantly greater than the distance between the atoms. Even if the intricate details of atomic interactions are not directly observed, this approach allows for a comprehensive understanding of the behavior of the material. This classic electrodynamic method was discovered long before the fundamental atomic structure of matter was understood. In this approach, we treat the material as a homogeneous medium, defined by its permeability ($\mu$) and permittivity ($\varepsilon$). These fundamental properties play a crucial role in determining how the material interacts with external magnetic and electric fields. The theory of homogenization is particularly effective across a wide range of wavelengths within the EM spectrum. However, as we delve into scales that are close to atomic dimensions, the limitations of this method become apparent. At these minute scales, the arrangement and behavior of the matter are heavily influenced by thermodynamic principles, reflecting the complex interplay of forces that govern matter on a scale only slightly larger than the atomic level. In recent years, there has been a burgeoning interest in exploring the EM characteristics of synthetic materials containing "atom-like" inclusions designed to interact with incoming waves in unconventional manners, resulting in unattainable properties with natural substances. The term 'Metamaterial' was first coined by Rodger M. Walser in 1999. Metamaterials are macroscopic composites made of periodic or non-periodic structures, and their functionality is determined by both the cellular architecture and chemical composition [4].

### 1.2 Why meta?

The question comes, "Why Meta?"; the word 'Meta' has a Greek origin meaning 'beyond' or 'after.' So obviously, MMs are materials that show characteristics beyond natural materials. They are made with capabilities that are extraordinary and are fed with abilities to manipulate waves that are not possible with natural materials from the periodic table [5]. As we know, in the domain of materials science, pure substances are characterized by a homogeneous chemical composition, such as metals or semiconductors, whereas mixtures are assemblies of substances that maintain their individual properties. However, metamaterials blur this categorization [6], as they are frequently engineered as structured amalgamations of constituents (e.g., metals and dielectrics) that are combined to yield emergent properties. Now, when considering the metamaterial as an effective medium, it becomes necessary for the size of the cells to be either smaller than or equal to the wavelength of the incident radiation to exhibit the required characteristics, and that refers to its sub-wavelength features [7], where the sub-wavelength features denote the structural components of a metamaterial that are significantly smaller than the wavelength of the EM wave that interacts with the medium. This condition guarantees that the EM wave interprets the metamaterial as a homogeneous medium rather than as an assemblage of discrete elements [6]. If '$a$' is the lattice and '$\lambda$' is the wavelength of the incoming radiation, it must satisfy the condition that,

$$a \ll \lambda \tag{1}$$

This helps in homogenization and tuning material response. Here, the equivalence of meta-atomic [8] structures with specific geometry and design can exhibit the same EM properties that are desired. The design flexibility in shape, size, and materials results in achieving emergent characteristics such as negative refractive index, unusual permeability, and permittivity values [9]. These materials have remarkable optical properties, like absorption, shielding, cloaking, and negative refraction.

### 1.3 The history behind the left-handed materials

The term "Meta" within the realm of a research subject concerning invisibility and "metamaterials" was initially brought into prominence by scholars in EM theory and optical engineering. The emergence of this concept gained widespread acknowledgment following the release of two seminal papers in 2006. The first one introduced to Ref. [10] entails directing EM waves around an entity to render it imperceptible. The second presented an alternative strategy employing conformal mapping [11] techniques to achieve invisibility using metamaterials. These materials are characterized by a dual negative index, which makes them left-handed materials (LHM) in the $\varepsilon - \mu$ domain [12] (the four quadrants of isotropic materials including double negative (DNG), Double positive (DPS), Epsilon negative (ENG), and μ-negative materials (MNG) [5]) a challenging feat to replicate in nature. Consequently, the approach involves artificially engineering these materials to possess the desired properties, but how was this done? The breakthrough in achieving this feat dates back to 1996 and 1999 with the invention of the split ring resonator (SRR) composed of both electric [13] and magnetic plasma [14]. Thus, it was understood that individual resonant structures of different sizes (wavelength range), shapes, designs, and patterns constitute bulk 3D MMs of desired behavior. For DNG materials, n also becomes negative. If the real part of ε and μ is negative, the refractive index will be positive as $n = \sqrt{\mu\varepsilon}$. However, energy loss leads to an imaginary component in both permittivity and permeability, so that the real part of n in the complex plane is negative. For materials with $n < 0$, the wave vector (k) and Poynting vector point in opposite directions, while the parallel component of the wave vector remains the same (**Figure 2**) [15].

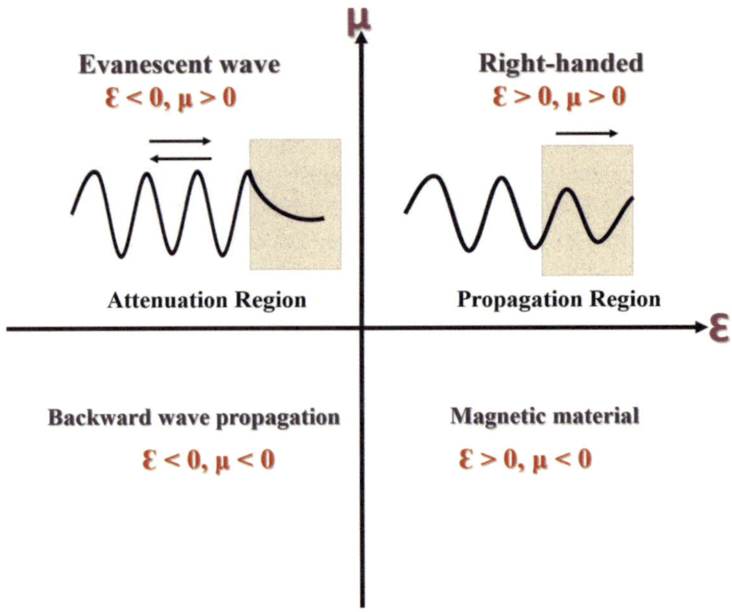


**Figure 2.**
*Constitutive parameters classification.*

In conventional materials with a positive n, the electric field (E), magnetic field (H), and k form a right-handed coordinate system. On the other hand, − n materials form a left-handed coordinate system. This chapter delves into the intriguing applications of metamaterials, a class of engineered materials with unique properties. We will begin by exploring electromagnetic (EM) shielding, which effectively protects sensitive electronic equipment from interference. Next, we will examine the role of metamaterial absorbers, which are designed to capture and dissipate electromagnetic waves across a range of frequencies with remarkable efficiency. Finally, we will discuss cloaking technologies, which aim to render objects invisible to detection by manipulating how light and other waves interact with them. Although all these applications leverage the principles of metamaterials, the methodologies employed to achieve each result are distinctly varied and fascinating.

## 2. Electromagnetic interference (EMI)

EMI poses a significant challenge in modern electronics, communication systems, and industrial applications. As electronic devices become increasingly compact and interconnected, their susceptibility to unwanted radiation has increased, resulting in degradation and system failure. Electrostatic discharge is a major source of EMI. It can cause signal interruptions, such as visual distortions on television screens or abrupt flashes, and aural disturbances, such as unexpected pops in audio systems. The two primary modes of existence of EMI between a source and a target device are conduction and radiation. The conduction means that an unwanted electromagnetic signal is imparted via physical connections; these include power or signal cables. When there is an electrical link between the EMI source and the receptor device, this interference may hop across the pathways, affecting the proper functioning of the receptor. Radiation has also been called the emission of EM waves from a source, which propagates in the surrounding medium and then impacts the receptor device [16]. These two forms of interference do not often exist independently; radiation and conduction are often present together. In practical situations, both need to be treated conjunctively; otherwise, any solution will be ineffective, particularly in electronic systems related to high-frequency signal processing.

### 2.1 EMI shielding

EMI shielding refers to the practice of enclosing electronic components or systems with materials designed to block external EMI. EMI can disrupt or degrade the functionality of sensitive electronic devices, posing significant challenges in fields ranging from telecommunications to medical devices. The early twentieth century conductive enclosures meant to diminish radio frequency interference (RFI) in communication systems. Conductive metals such as copper and aluminum were used as the early shielding materials for their highly conductive features and ease of fabrication [17]. In modern shielding strategies, some combinations of conductive coatings, metal enclosures, ferrite beads, along with the latest materials to attenuate conductive and radiative coupling routes, are employed within EMI control strategies. These methods have been known to be a major part of the communications, aerospace, automotive, and consumer electronics industries, where integrity of the signal is key, and their understanding of EMI in both its modalities helps engineers design systems more resistant to EM disturbances. Materials such as composite and other non-metallic shielding solutions would provide better performance in certain applications. Carbon-based polymer and ferrite materials emerged as alternatives to metal, which are excellent conductors and can reflect or absorb EMI when employed as EMI shield.

There is rapid increase in the development of shielding standards; the IEC and FCC recently adopted standards detailing the parameters for shielding effectiveness (SE) and compliance benchmarks for different industries. Modern regulations impose a high degree of constraints, in addition to shielding efficiency, strong emphasis on the environmental aspects at large, such as recyclability, and diminished EM pollution [18].

### 2.2 Mechanism of shielding

The major mechanisms of shielding from EMI include absorption, reflection, and multiple reflections. Absorption occurs when EM waves penetrate a shielding material and dissipate some of their energy as heat through resistive and dielectric losses within the material [19]. This mechanism is very effective for materials with high electrical conductivity and high magnetic permeability, such as steel and nickel. Reflection, on the other hand, is charged up surfaces generating opposing currents in response to the incident EM wave, resulting in wave reflection. This mechanism has a very strong influence on the electrical and magnetic values of the shielding material and the frequency of incoming waves. Also, there is a possibility of multiple reflections whereby waves will bounce numerous times internally within the thin, porous layers, leading to cumulative attenuation in the propagation of EM field energy. Nevertheless, performance specifics of the EM shielding are dependent upon materials' properties: conduction ($\sigma$), permeability ($\mu$), and permittivity ($\varepsilon$). For example, copper materials are known for high conductivity and hence great reflection from high-frequency electromagnetic fields; mu-metal, Meta surfaces afford strong shielding inserts against lower-frequency magnetism [20].

### 2.3 Shielding effectiveness (SE)

SE describes a material's capability to attenuate or block EM waves, using decibels (dB) as units. It incorporates three core components, which include reflection (R), absorption (A), and multiple reflections ($M$). The total shielding effectiveness is given by:

$$SE = R + A + M \tag{2}$$

*Reflection loss (R)* is due to the impedance mismatch created between the incident waves and the shielding material. When the EM waves impinge on a medium of impedance dissimilar to that of free space ($Z_0 \approx 377\ \Omega$), some of the energy will reflect back. The reflection loss can be calculated using the expression:

$$R = 20\,log\ \log_{10}\left(\frac{Z_0}{Z_S}\right) \tag{3}$$

Where $Z_S$ is the impedance of the shielding material defined as:

$$Z_S = \sqrt{\frac{\mu}{\sigma}} \tag{4}$$

Here, $\mu$ is the permeability, while $\sigma$ represents the electrical conductivity of the material. The higher the mismatch in impedances, the higher the reflections [21].

*Absorption loss (A)* occurs when EM wave energy is converted to heat as it travels through a shielding material owing to resistive and dielectric losses. The absorption loss is proportional to the material's thickness (t) and the wave penetration depth as described by the skin-depth ($\delta$), which signifies how deep a wave can penetrate a conductor before it attenuates to 1/e of its initial amplitude shown in **Figure 3**. The skin depth is given by,

$$\delta = \sqrt{\frac{2}{\mu\sigma\omega}} \tag{5}$$

Here, $\omega$ is the angular frequency of the wave. It is because of high absorption loss with reduced skin depth that high-conductivity and permeability of materials like copper or steel are particularly effective at shielding. The absorption is given as:

$$A = \frac{8.686 \times t}{\delta} \tag{6}$$

It thus shows that thicker materials give larger absorption loss (A). With increase in frequency, the skin depth decreases, thereby causing current to be made to flow closer to the surface of the shielding material.

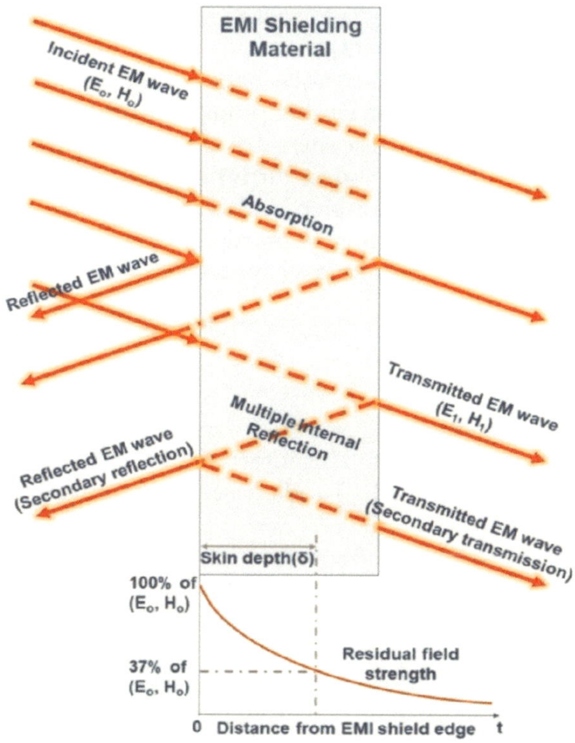


**Figure 3.**
*EMI shielding mechanism [22].*

This ensures that high frequencies are prevented from getting more access to the shielding material as much lower thickness is required to prevent them, and thicker sheltering systems are employed for low frequencies or the deployment of more permeable shielding alloys. Multiple reflections (M) are, however, of lesser importance and contribute to attenuation where multilayered, porous, or granular materials cause EM waves to bounce internally multiple times [22].

## 2.4 Theoretical foundation of shielding

The behavior of EM waves in shielding materials is regulated by Maxwell's equations, which explain the relationship between electric (E) and magnetic (*H*) fields. For time-varying fields, the equations become:

$$\nabla \times E = -\frac{\partial B}{\partial t} \left(\text{Faraday's Law}\right) \tag{7}$$

$$\nabla \times H = J + \frac{\partial D}{\partial t} \,(\text{Ampere} - \text{Maxwell Law}) \tag{8}$$

Where $B = \mu H$ and $D = \varepsilon B$; relate magnetic and electric fields to material properties, including permeability ($\mu$) and permittivity (ε). Materials of high conductivity ($\varepsilon$) are excellent reflectors and are very effective against high-frequency waves. On the other hand, materials of high permeability ($\mu$), such as mu-metal, are more effective in low-frequency magnetic field shielding [23].

Microwave-absorbing materials are noble and of utmost consideration in radar stealth technologies, communication system protection, and electromagnetic pollution control. Their capacity to convert energy of electromagnetic waves into heat or dissipate it as dielectric or magnetic losses turns them into essential materials considering worldwide engineering applications. Mathematically understood, the Maxwell-Wagner effect is related to the complex permittivity ($\varepsilon^*$) of a material, any material that would possess energy storage and loss capabilities. With respect to this, it can be expressed as

$$\epsilon^* = \epsilon' - j\epsilon'' \tag{9}$$

where $\varepsilon'$ is the real part representing energy storage capacity, and $\varepsilon''$ is the imaginary part associated with dielectric losses. This interfacial polarization is accordingly believed to increase the loss factor ($\varepsilon''$) and consequently enhances microwave absorption.

Within the limits of effective medium theories, such as Maxwell-Garnett and Bruggeman models, explanations of the mechanism predicting the composite permittivity fix the level of interfacial polarization. These explained how the respective dielectric properties of their components determine the overall behavior of heterogeneous materials. By regulating the distribution and concentration of conductive fillers, interfacial polarization effects can be enhanced in microwave absorption [23, 24]. The combination of porous architecture, conductive nanostructures, and dielectric matrices has been paramount to the development of next-generation materials for telecommunications, defense, and electronics applications.

### 2.5 Materials being used for EMI shielding

Graphene and its derivatives—RGO genre—are gradually being looked at with favor for EMI shielding owing to their excellent conductivity and extremely high surface-to-volume ratio. Wave reflection can significantly take place because of graphene being an excellent conductor, while fine, thin, and flexible sheets provide lightweight protection for shielding. Combined with SF and PMMA, the RGO has shown to make a potent character in microwave absorbers as developed by Acharya et al. Subsequently, this produced a conductive network providing both dielectric and magnetic loss mechanisms. The optimized composite with 9 wt% RGO and 1 wt% SF offered absorption efficiencies exceeding 99% in the X-band (8–12 GHz) through magneto-dielectric coupling [24]. The Maxwell-Wagner effect plays a critical role in the absorption of microwaves by considering interfacial polarization in heterogeneous dielectric materials. This appears when an alternating EM field encounters a composite matured by components having different electrical properties (i.e., different $\sigma$ and $\epsilon$). When electromagnetic waves are penetrating a composite medium either thick or thin, there is a charge build-up occurring at the interfaces of materials with conductors and insulators as they do not instantaneously respond to the external field, which significantly varies with time. Due to slow motion of the charge, this relaxes fast enough to permit interaction with the quick field changing, which gives rise to localized polarization. This charge polarization dissipates energy from the incident wave, thus adding to the microwave absorption efficiency of such materials, which is much augmented in materials combining conducting nano supports such as CNTs,

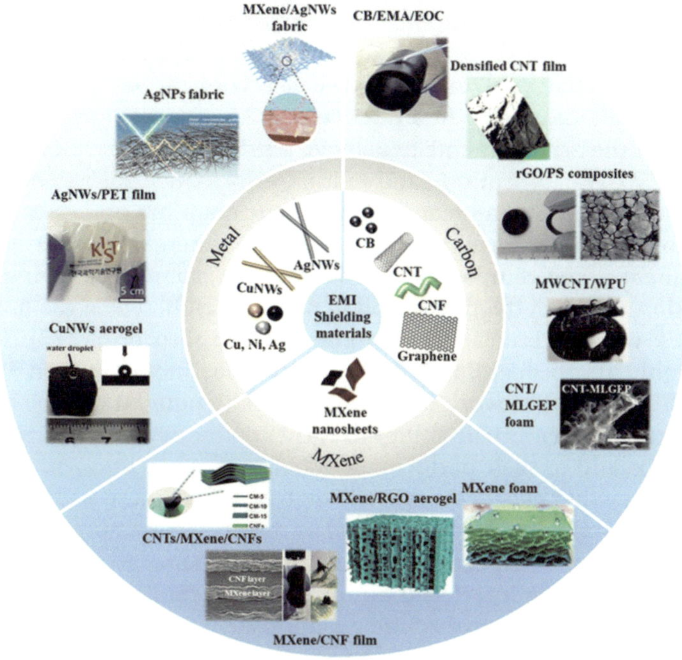


**Figure 4.**
*Illustration of different materials used for EMI shielding [19].*

MXene, and graphene-with dielectric matrices such as polymers, aerogels, or nanocellulose, as shown in **Figure 4**. This results in innumerable interfaces where charge builds up and realigns via microwaves. Nanocarbon-structured aerogels combined with CNTs showed far superior absorption power by capturing and dissipating EM waves because of the additional contributions from polarization effects [25].

This interfacial polarization enables attenuation of the energy and dipole formation to further hamper the propagation of an electromagnetic wave. The composites built on this polarizing mechanism derive reinforcement through synergistic interfacial interaction from conducting fillers into insulating matrices. The MXene-polymer composites indicated innovations in dielectric properties, essentially having complementary needs for interfacial charge storage and dissipation. In that regard, Mendoza-Sánchez et al. cited that the performances would be phenomenal with harnessing the maximization of the parameter relating to texture and chemistry-based assemblies for otherwise crystallization in nano-dimensional fortified crystals. Its dispersed phase lengthened the path for microwave propagation through enhanced interfacial zones because of the porous architecture of high-open-cell anti-microbial advanced materials. They also make a significant contribution toward the scattering processes, which build up localized charges at interfaces [26]. Zhou et al. demonstrated that well-connected, dense sheet-like networks from carbon aerogels showed excellent absorption capabilities working at X-band frequencies due to their interfacial polarization capabilities [27].

Carbon nanomaterials, such as carbon black, multi-walled carbon nanotubes (MWCNTs), graphene, and graphene nanoribbons (GNRs), have also demonstrated their efficiency for EMI shielding. Tripathi et al. studied the absorption of microwave-perfect MWCNT and carbon black-reinforced lightweight polyurethane (PU) matrix nanocomposites. The optimized sample with a particular content of 75 mg MWCNT and 250 mg carbon black per gram of PU demonstrated a peak reflection loss of 16.14 dB at 11.92 GHz with about 90% absorption over the 11.03–13.00 GHz frequency range [28]. Yadav et al. examined the effects of carbon black-based composites containing cobalt and $MoS_2$. At a thickness of 2.5 mm, the incorporation of such materials allowed the arrangements to achieve a shielding efficacy of over 97% at 14 GHz with a 7.7 GHz absorption bandwidth. Performance absorption was enhanced by asymmetric polarization and magneto-dielectric coupling processes. This work reports high absorption performance because of the synergistic effect of conductive filler enhancement of the complex permittivity of the material, allowing for enhanced wave attenuation and the effective coupling mechanism between the magnetic and dielectric losses, which is crucial for broad microwave absorption [29]. The composite presented roughly 40 dB shielding effectiveness, reflecting more than 95% relative to the incoming EM wave. The work of Alegaonkar et al. pointed to edge effects of the GNRs, which, coupled with their ability to form conductively connecting clusters within the matrix, add to their performance shall be flashed into a braze by the conductively connecting clusters inside the matrix. GNRs increased the matrix's dielectric and conductive properties, as well as the efficiency of microwave absorption. The polyurethanes-composite incorporated with nanocarbons showed improved absorption owing to strong interfacial interactions and multiple scattering mechanisms. Using hierarchical structures into a composite allowed one to trap electromagnetic waves by extending the path of its propagation toward relaxation into dielectric and conductivity losses. The underlying mechanisms that govern the microwave absorption processes in these nanocomposite systems are of a complex nature. Dielectric losses due to phenomena, such as dipole and interfacial polarization, are vital for energy dissipation. Besides, magnetic loss mechanisms due to natural and exchange resonance give up their contribution to absorption once magnetic inclusions occur, for instance, with cobalt nanoparticles. Eddy current losses, due to circular current movement within ferromagnetic materials, give additional channels for heat dissipation. The losses due to conduction are another major aspect of microwave absorption [30].

Graphene nanoribbons are another potential option for EM shielding owing to their architecture and larger aspect ratios. Joshi et al. synthesized a hybrid composite consisting of GNRs, polyaniline (PANI), and epoxy resin, showing a shielding effectiveness greater at X-band. Materials such as carbon black and MWCNTs organize themselves into conductive paths through which EM energy can be directly converted to heat via the motion of electrons. Photothermal conversion properties of materials like MXenes allow this to be much more pronounced and would be extremely beneficial for broadband absorption [31]. Added to this is the hierarchical porous structure, such as that found in nanocarbon aerogels and composites, which enables multiple scattering of electromagnetic waves. Providing an extended path for wave propagation enhances absorption efficiency by capturing and distributing wave energy. Analysis of EMI shielding characteristics of hybrid composites manufactured using a large number of polymer hosts by incorporating a variety of fillers such as graphene, graphene derivatives, magnetic nanoparticles, conductive polymers to obtain shield with optimal efficiency and payload has been reported extensively. Between absorption efficiency and density, this question remains largely unanswered; however, derivatives and other modalities comes off as very promising in its scope of application range covering domains from stealth technologies to defense systems and devices of commercial electronics.

## 3. Metasurface

While metamaterials have captured our imagination by manipulating EM waves in desired ways, their practical application is sometimes hindered by issues including bulkiness, inherent loss, and design-production complexity. This has resulted in the development of MSs, which are 2-D equivalents of metamaterials that preserve metamaterial properties and, simultaneously, offers benefits over them, such as lower loss values and design complexity [32]. Utilizing metasurface has garnered considerable interest due to the streamlined design, made possible by the generalized Snell's principles governing reflection and refraction. It was demonstrated that manipulation of wave propagation is achievable by implementing a thinly coated layer with a meticulously crafted phase gradient across the surface [33]. This sub-wavelength design allows for unprecedented manipulation of light waves, enabling a range of extraordinary optical phenomena and capabilities. The central ambition behind the development of metasurface is to unlock enhanced control over light, paving the way for advancements in various applications, from cloaking devices to superlenses and beyond. Over the past 20 years, metasurfaces have progressed rapidly. One of the most challenging aspects is the realization of these metasurfaces. Various structures can often be utilized to create metasurfaces, including photonic crystal structures, transmission line-based designs, resonant structures, and fully dielectric materials.

a. *Photonic crystals*: Dielectric metals, superconductors, and semiconductor microstructures are intricately organized in regular patterns across one, two, and three dimensions to form what are known as photonic crystals. These structured arrangements are designed to manipulate the behavior of light by creating a forbidden gap in the dispersion band diagram. This gap plays an important role in controlling the propagation of EM waves, allowing for precise regulation of their movement and interaction within various applications in photonics and optoelectronics.

b. *Transmission line (TL) structure*: By incorporating shunt inductance and series capacitance into a standard printed circuit board transmission line, it is possible to create a negative phase velocity regime.

c. *Resonant structures*: Resonant structures are characterized by both magnetic and electric resonance. The electric resonance generates a -μ due to the induced B created by a current loop in response to an incident B that is below the electric plasma resonance frequency. Similarly, an electric resonant structure can produce negative permittivity using the same principle. As a result, it is possible to achieve both -ε and -μ simultaneously by combining electric and magnetic resonance.

In addition to the advancements in metasurface, researchers have predicted and demonstrated a diverse array of potential applications in the past decade. These include innovative technologies such as cloaking, EM wave absorbers, energy harvesting systems, telecommunications, medicine, and energy [34].

### 3.1 Metasurface as EM absorber

The modern world is profoundly reliant on a vast array of technologies that harness EM radiation, a fundamental force of nature that underpins many aspects of our daily lives. Pioneering physicist James Clerk Maxwell made groundbreaking contributions to the study of electromagnetism in the nineteenth century, laying the groundwork for numerous innovations that have transformed modern society. His work led to the creation of radio, television, wireless communication, global navigation, radar systems, and other technologies that support our rapidly evolving civilization [35]. One important aspect of aeronautical engineering is EM absorption, which plays a crucial role in stealth technology. The goal is to create a thin, lightweight material that effectively absorbs a wide range of frequencies. When EM energy from radar is absorbed by this material, the amount of energy reflected back decreases. As a result, an aircraft can become less detectable by radar by utilizing materials that are more effective at absorbing EM waves. Microwave absorbers are highly sought after because of their applications in both the military and civilian sectors. In the defense industry, these absorbers serve covert purposes, such as reducing radar cross-section (RCS) and providing protection against microwave radiation. A specific area within electronic warfare countermeasures, known as "stealth technology," employs various techniques to make military systems such as tanks, battleships, missiles, and aircraft less detectable, or even invisible, to enemy radar. To reduce the RCS of an object, its external features can be shaped to minimize the backscattering of EM waves toward the radar source. Additionally, coating the target surface with radar-absorbing materials can also lower the RCS [36]. The increasing use of electronic devices such as computer local area networks, ovens, palmtops, medical equipment, Wi-Fi systems, and satellite communications has led to a greater focus on microwave absorbers in the civil sector. These devices generate more EM noise each day, which can result in cross-talk and contribute to environmental pollution from microwave radiation. To address these issues, EM compatibility (EMC) and the use of microwave absorbers are essential [37].

### 3.2 Progress in EM wave absorber

All incident electromagnetic waves, regardless of their frequency, polarization, or direction, can be absorbed by a perfect wave absorber. This type of absorber does not produce any reflected waves. However, a truly perfect EM wave absorber does not exist in reality. As a result, the effectiveness of material that absorbs EM waves is evaluated based on predetermined acceptable values for reflection coefficients. Typically, reflection coefficients of −20 dB or less are used as standards, while for applications requiring high performance, a coefficient of −30 dB or less is expected. A reflection coefficient of −20 dB indicates that the wave absorber absorbs 99% of the incident electromagnetic wave energy, while −30 dB indicates 99.9% [38].

Due to the high demand for wave absorbers in military applications, research began during World War II. Two different kinds of absorbers were architected as part of the German Schornsteinfeger Project and were mounted on the periscope and snorkel of a submarine to help conceal radar signals [39]. One of them was made from a material called Wesch, which consists of a rubber sheet infused with carbonyl iron. The second type, known as the Jaumann absorber, was created by alternating layers of dielectric material and a resistance sheet [40]. Additionally, the Salisbury screen absorber has been proposed [41]. The structure of this resonant-type wave absorber was made up of a resistance sheet that had a resistivity of 377 Ω and was positioned $\lambda/4$ from the rear metal plate. **Figure 5** shows the design of the proposed absorber.

In the realm of metamaterials, specific frequency bands have been deliberately targeted in the design of conductive absorbers [43]. However, the pursuit of wide-band or broadband absorption can impose limitations on the selectivity of these

*DOI: http://dx.doi.org/10.5772/intechopen.1009940*

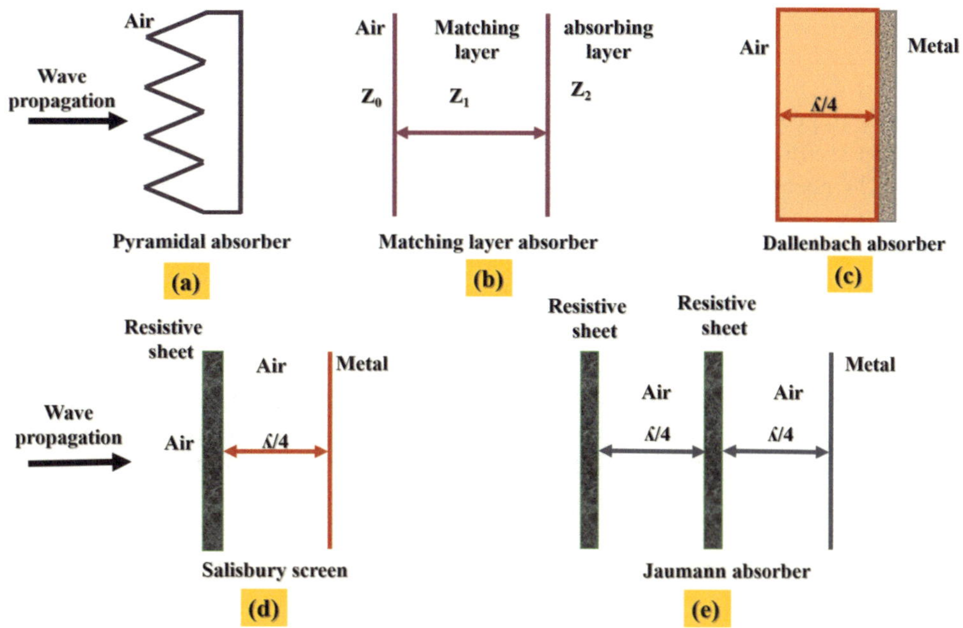


**Figure 5.**
*Design of absorbers (a) pyramidal absorber, (b) Matching layer absorber, (c) Dallenbach absorber, (d) Salisbury absorber, and (e) Jaumann absorber [42].*

frequency bands. Numerous absorbers have been developed that enhance the bandwidth of absorption, including innovative designs such as resistive sheets and lumped resistors. While these solutions can significantly improve performance, incorporating many lumped resistors often leads to increased manufacturing costs and adds complexity to the production process [44]. An intriguing alternative to traditional methods is the use of resistive ink, which can be applied to dielectric surfaces to achieve broad absorption bands. Although resistive paint tends to be more cost-effective, its successful application demands precise control over a uniformly thick dielectric base. Fortunately, modern processing techniques like screen printing and economical liquid-jet printing make it easier to achieve this precision, ensuring that the resistive ink is evenly distributed for optimal performance [45–47].

### 3.3 Mechanism of EM/ microwave absorber

#### *3.3.1 Minimization of reflection*

In the context of metal-backed single-layer absorbers, the phenomenon of reflection is primarily observed at two key locations: the interface where the air meets the absorber and the metal backing that supports the absorber. This reflection can lead to significant losses in the effectiveness of EM absorption. To reduce these reflections at the interface, it is crucial to align the input impedance of the absorptive layer as closely as possible to that of free space. Achieving this match helps to ensure that EM waves are allowed to penetrate the absorber rather than bounce back into the environment. To facilitate this process, a specific formula is employed to find the normalized input impedance (Zin) at the air-absorber interface in relation to free space impedance ($Z_0$). This calculation is essential for optimizing the performance of the absorber [48].

$$Z_{in} = \sqrt{\frac{\mu_r}{\varepsilon_r}} \tanh\left[\left(-\frac{j2\pi f}{c}\right)\sqrt{\mu_r \varepsilon_r t}\right] \tag{10}$$

Where c is the velocity of the wave, t is the width of the absorber, f is the incident microwave frequency, $\mu_r = \mu_r' - \mu_r''$, and $\varepsilon_r = \varepsilon_r' - \varepsilon_r''$. For effective impedance matching between the absorber and the free space interface, the ratio of the real μ to ε should be close to unity. In dielectric composites, the intrinsic adjustable parameter, μr = 1–j.0, is used to align the real part of permittivity with unity, thereby achieving optimal impedance matching.

*3.3.2 Thickness of the absorber*

By reducing the reflection from both the front and rare interfaces, phase cancelation can be employed to minimize the overall reflected wave. This process follows the principle of destructive interference, which can be described by the equation as:

$$t = \frac{\lambda_0}{4\sqrt{|\mu_r||\varepsilon_r|}} \tag{11}$$

Where t is the width of the absorber, and $\lambda_0$ is the wavelength in free space.

*3.3.3 Reflection loss*

From the transmission line theory (TL), the reflection loss of the absorber can be expressed as [49]:

$$R_L = -20\ log_{10}\left[\left|\frac{Z_{in} - Z_0}{Z_{in} + Z_0}\right|\right] \tag{12}$$

Where, $Z_0$ is called the free space impedance and is equal to 377 Ω. For minimum reflection loss, $Z_{in}$ should be equal to $Z_0$.

## 4. Application of microwave absorber

Microwave absorbers are specialized materials characterized by their exceptional properties, such as high dielectric loss and enhanced magnetic permeability. These attributes enable them to effectively dissipate microwave radiation energy, which is crucial in minimizing interference and optimizing overall system performance. As a result, they play a vital role in improving energy efficiency across a diverse range of industries. The applications of microwave absorbers are numerous and varied, including radar and stealth technology, antenna design, microwave oven, medical applications (thermal therapy), environmental and industrial monitoring, and communication systems.

## 5. Cloak-innovation

What if we could truly become invisible? How would this impact our interactions, understanding of privacy, and relationship with the world around us? While this might seem like a fantastical notion, scientists are actively working to make invisibility a reality. It was in 2006 that the first breakthrough occurred in the field of invisibility and non-linear optics. Sir John Pendry proposed the idea of a 'meta'-material cloak that could attain invisibility in at least some frequency range [10]. Through their work, they demonstrated a sort of material that has a significant advantage in emphasizing the material parameters that are non-existent by altering the structural design. Yes, traces of invisibility are possible due to continuous research and scientific findings in the novel era.

Fueled by breakthroughs in material science and engineering, the field of cloaking technology has witnessed considerable progression in recent years. Scholars are investigating sophisticated materials endowed with distinctive properties absent in the natural world that can manipulate EM waves to facilitate invisibility. Innovative design methodologies employing methods such as transformation optics (TO) are allowing for enhanced precision in wave propagation control, thereby increasing the efficacy of cloaking devices across a wider range of frequencies. Present applications extend beyond mere optical cloaking to encompass acoustic cloaking for sound attenuation and thermal cloaks designed for the regulation of heat transfer. These advancements are unlocking opportunities for application across diverse sectors, rendering cloaking technology a multidisciplinary frontier of innovation.

### 5.1 Evolutions and foundations of cloaking technology

An EM cloak fundamentally operates as a device that induces a state of "invisibility" for an object concerning EM radiation within a designated frequency range [50]. The concept of a device that can render objects imperceptible to the unaided eye is deeply entrenched in a rich historical narrative prevalent in numerous cultures' folklore, including the renowned cloak associated with Harry Potter from the literary works of J. K. Rowling [50]. Although early researchers were unable to attain genuine invisibility, their investigations catalyzed modern advancements in metamaterials and cloaking technologies, wherein the manipulation of light facilitates the perceived "disappearance" of objects by redirecting light around them. Nonetheless, the viability of attaining genuine invisibility, albeit within a limited spectrum of frequencies [1], has persistently been a focal point of investigation. Several scholarly works, including those by Leonhardtand [11] Pendry [51, 52], explored the foundational principles of invisibility mechanisms for objects. The inception of a novel domain emerged, integrating linear and non-linear optical concepts encompassing various principles from optics and wave analysis, such as Snell's laws [52], Fermat's principle, and so on [53]. An entity is rendered undetectable when it does not reflect waves to the source, refrains from dispersing waves through alternate pathways, and avoids generating any shadow. In other words, the entity must not disrupt the surrounding fields, or the fields should propagate undisturbed such that no object obstructs their path [50]. In EM wave scattering theory, the notion of "cloaking" pertains to the attenuation of an object's overall scattering cross section (SCS) [50, 54] to ideally reduce it to zero.

$$SCS = \frac{Total\ scattered\ power}{Incident\ power} \tag{13}$$

Where SCS is the scattering cross-section. Accomplishing this entails either surrounding the object with an absorbing medium or altering its geometry to diminish the scattering field aimed toward the illumination source. Optimal cloaking encompasses the potential to render the entire object imperceptible, with both back-scattering and shadow being completely or partially mitigated. This accounts for the background of the cloaking technology.

### 5.2 Foundational concepts

Each type of wave interacts with matter through mechanisms such as absorption, reflection, refraction, and scattering, making specific regions of the spectrum crucial for tailoring cloaking technologies. Longer wavelengths, including radio and microwaves, are extensively utilized within stealth technologies [55]. Infrared cloaking is centered on the reduction of thermal signatures, a critical capability within military surveillance frameworks to evade thermal imaging techniques [56]. In the visible spectrum, the challenge of cloaking becomes most pronounced, necessitating meticulous manipulation of light scattering and refraction to achieve invisibility to the human eye, leading to intricate lambda-d correlation issues [57]. By leveraging advanced materials like metamaterials, which help to manipulate and control EM waves in extraordinary manners, cloaking technologies can be tailored to function across designated wavelengths, thereby transmuting concepts that were once relegated to the realm of science fiction into tangible reality.

### 5.3 RC delays v/s sub-diffraction limit

As we investigate the practical execution of cloaking devices more thoroughly, an additional pivotal factor emerges—the "temporal" response of these materials to EM waves. This response is shaped by inherent characteristics such as resistance and capacitance [58], which together contribute to RC delays [5]. While optics is primarily concerned with the manipulation of spatial wave propagation, RC delays underscore the significance of overseeing time-dependent behavior within synthetic materials. Comprehending and mitigating these delays are essential for the proficient real-time functionality of cloaks, especially when dynamic or broadband performance is needed [59]. RC delay refers to the time constant indicated by,

$$\tau = R \times C \tag{14}$$

Where R is the resistance, and C is the capacitance.

It should be understood that high RC delays will slow response, limiting the cloaking device's operating bandwidth and creating energy loss. As a result, cloaking structures should be designed in such a way that minimizes the RC delay values within the 'sub-diffraction limit' [60]. Sub-diffraction control of light waves is necessary for the seamless bending and propagation of light. However, incorporating minimum RC delay and residing within the sub-diffraction limit seems challenging for the cloaking to be in effect [61]. More advanced materials with low-loss characteristics, high conductivity, and low resistance can minimize RC delays, while advanced nanofabrication techniques for the designing of novel materials to enable sub-diffraction precision are also in consideration for higher frequencies involving visible light.

### 5.4 Core principles of cloaking

The Maxwell equations are four mathematical equations delineating the characteristics of EM waves. Maxwell's equations for plane waves can be written as,

$$K \times E = \omega\mu H \tag{15}$$

$$K \times H = -\omega\varepsilon E \tag{16}$$

But for the left-handed metamaterials or DNG materials,

$$K \times E = -\omega\mu H \tag{17}$$

$$K \times H = \omega\varepsilon E \tag{18}$$

Here, the light propagation will be opposite to what happens in natural materials.

*Snell's Law*: Using Snell's Law is essential in metamaterial engineering, particularly in cloaking. It is common knowledge that materials in the third quadrant of the material profile exhibit a negative refractive index [62]. Consequently, when an incident ray interacts with such materials, it experiences negative refraction or diffraction [63]. The refractive index of metamaterials will be thus;

$$n = -\sqrt{\mu\varepsilon} \tag{19}$$

Where $\mu$ is the permeability, and $\varepsilon$ is the permittivity of the medium.

### 5.5 Development of cloaking technology

As mentioned, the root of cloaking technology traces back to the early twenty-first century, driven by advancements in EM theory and material science. The practical demonstrations of novel methods were truly influenced and initiated by the TO method and the advent of metamaterials. Over time, researchers focused on more efficient methods for the attainment of less bulky, less-lossy, high-performance cloaks. They refined these principles and incorporated new ideas, achieving breakthroughs in various frequency ranges, from microwaves to visible light. With the advent of these technological advancements, a diverse array of cloaking methodologies has emerged, employing distinctive techniques for the manipulation of interacting waves.

#### *5.5.1 Modern cloaking technologies*

Certain cloaking techniques encompass SRR-based cloaking devices, which have gained widespread application to shield objects at microwave ranges [15, 64]. Transmission line cloaks [65] incorporate engineered transmission lines, which function akin to artificial materials. Optical transformation cloaking, which utilizes principles of TO, is instrumental in redirecting waves along predetermined trajectories [11, 66]. Metamaterial cloaking, as a previously elucidated technique, exploits artificially engineered materials [10, 50]. Antenna cloaking serves to minimize the electromagnetic scattering induced by antennas. The scattering cancelation approach entails the application of coatings composed of materials designed to negate scattered waves [58, 67]. Furthermore, carpet method-based cloaking conceals objects by engineering the illusion of a flat surface [52, 68], which remains a significant focus of ongoing research in contemporary contexts.

#### *5.5.2 Hiding in plain sight: The carpet cloaking method*

Cloaking research has progressed substantially through the application of transformation optics (TO) and optical-conformal methodologies; however, these approaches encounter significant challenges when dealing with anisotropic materials, with high attenuation and dispersion effects [53]. To address these issues, the innovative concept of "hiding under a carpet" was proposed, wherein the reflection characteristics of a scatterer are strategically mapped [52]. This methodology effectively converts scattered waves into a planar reflection pattern, thereby generating a ghost image indistinguishable from the surrounding ground. By utilizing lightweight and compact metasurfaces designed with precise phase gradients, the scatterer can reflect waves in a manner that makes it appear as an integral part of the ground plane. This technique, which has been advanced by Li and Pendry, [52] facilitates seamless integration and offers practical solutions for achieving invisibility.

### *5.5.3 Research advancements*

Ongoing research activities show advancements in the field of carpet technique with more efficient performances. Herein we produce the efficient working of a carpet cloak made up of dielectric resonators and substrate with a one-order mismatch in their epsilon values. The simulations utilize the full wave solver, CST Microwave Studio Suite, using the Time Domain Solver within the 4–5 GHz frequency range. The excitation mode applied consists of plane waves exhibiting various polarizations, such as linear, circular, and elliptical. These waves were generated at a frequency of 4.15 GHz. To assess the similarity between the designed cloak-attached scatterer and the ground plane, E-field and far-field monitors were utilized at the frequency of 4.15 GHz. The results of the E-field distributions of plane-polarized waves are visually represented in the graphical illustrations presented in **Figure 6**. Here it is evident from the simulation results [69], the cloak attached to the surface minimizes the distortions caused by the scatterer, thus making it more similar to the pattern of a ground plane. It has been observed that the metasurface cloak can conceal the scatterer to a certain degree, thereby replicating the characteristics of the ground plane across all modes of plane wave excitation. The distortion, as mentioned earlier, introduced by the scatterer, is effectively mitigated by the same, resulting in a reflection profile akin to a quasi-plane wave [53]. Consequently, it is now apparent that discerning the scatterer from the ground plane is challenging for any observer. The reflection pattern of the electric field provides insight into the cloaking effect achieved by the metasurface [69].

Another significant outcome enhancing the cloaking efficiency is the Radar Cross Section (RCS), demonstrated by the relevant structures. The RCS values exhibited by the scatterer attached to the cloak are similar to those of the ground plane for various polarization scenarios, including circular and elliptical as our investigations [69], as shown in **Figure 7(a)**, **(b)**, and **7(c)**.

The carpet cloak presented compelling evidence of the manageability of EM waves through metamaterial technology and TO. Utilizing these non-resonant metamaterials

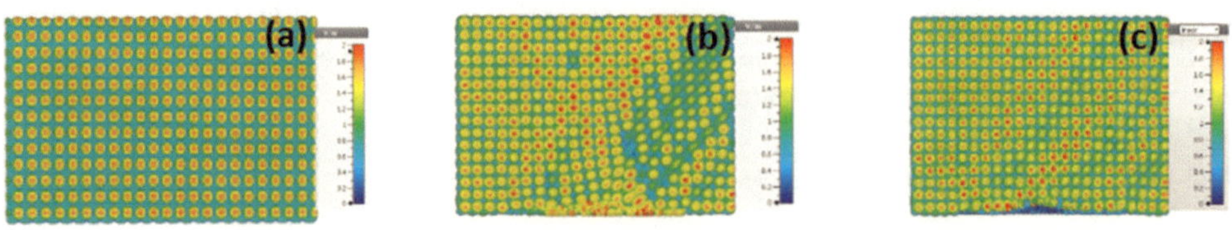


**Figure 6.**
*The electric field patterns (a) on a ground plane, (b) on a scatterer, and (c) on a cloak-attached scatterer.*

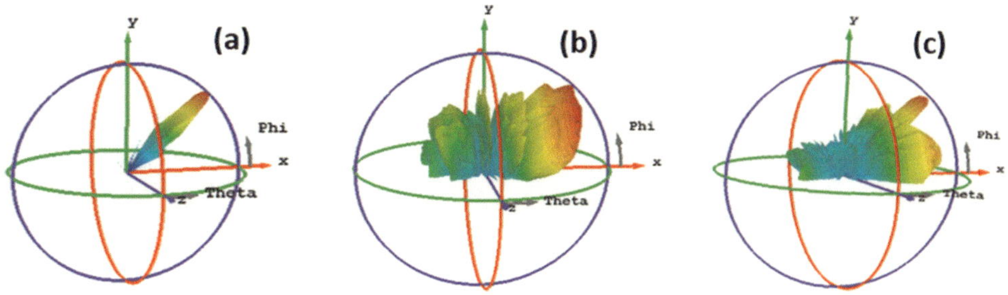


**Figure 7.**
*Simulation results of RCS values of the structures, (a) ground plane, (b) scatterer, and (c) scatterer attached with a cloak.*

can significantly enhance bandwidth and reduce loss performance. Despite being in the initial stages, the process of cloaking signifies a novel method to govern EM wave propagation through complex media, holding substantial potential for diverse RF and optical applications in the future.

### 5.6 Ongoing research and challenges

The field of cloaking expands by focusing on improving efficiency, scalability, and adaptability. Current research focuses more on broadband cloaking [10] apart from cloaking in some specific ranges or frequencies, [68]. Also, methods that would reduce material loss and anisotropic issues in the design of metamaterials are of concern [68]. The emergence of low-cost, versatile, less bulky, lightweight, and flexible cloaks is a major concern. Challenges also persist in dynamic or non-homogenous environments where the cloaks are exposed to changes in temperature, pressure, and so on. Research is now deepening roots in the fields of quantum cloaking and 4D metamaterials. But these advances should also pave the way for reducing the challenges that pertain because apart from theoretical predictions, our ultimate aim is to make the technology practically possible for humanity.

## 6. Conclusion

Metamaterials are gaining significant attention due to their unique physical properties. They can change electromagnetic wave behavior in a peculiar way that regular materials cannot. As a result, they can be a viable alternative in many engineering disciplines. Electromagnetic interference shielding is one of the important branches of communication engineering. Effective shields can conceal the incident radiation effectively. Hence, shielding effectiveness (in dB) is a crucial parameter in electromagnetic interference shielding architecture. The chapter presents state-of-the-art shielding strategies with the most recent relevant literature. Further, the absorption behavior of microwave absorbers has also been discussed in light of their design, mechanism of absorbance, and their ability to maintain absorption at larger angles of incidence. Meta absorber is thus a promising stealth technology. In cloaking, metamaterials are implemented to make objects almost invisible, which has further opened a new possibility in this sector. The goal in cloak is to create a thin, lightweight material that effectively absorbs a wide range of frequencies. This technology has evolved from mere a theoretical construct to a tangible application. Methodologies including transformation optical metasurface and scattering cancelation have established novel avenues for realizing invisibility. Nonetheless, challenges including material loss, anisotropy, and scalability underscore the necessity for ongoing investigation. The persistent quest for lightweight, less bulky, broadband, and dynamic cloaking solutions indicates a promising trajectory. As scientific inquiry advances, cloaking is set to transform industries, providing innovative remedies to practical challenges while fostering further inquiry into the limits of invisibility. So, yes, let us hope for the best in technology!

## Author details

Priyanka[1], Akshita Yadav[2], Aishwarya Pradeep[1], Himangshu B. Baskey[3] and Prashant S. Alegaonkar[1]*

1 Department of Physics, School of Basic Sciences, Central University of Punjab, Bathinda, Punjab, India

2 Department of Mechanical and Aerospace Engineering, State University of New York, Buffalo NY, USA

3 Stealth and Camouflage Division, Defence Material Store R & D Establishment, DRDO, Kanpur, Uttar Pradesh, India

*Address all correspondence to: prashant.alegaonkar@cup.edu.in

**IntechOpen**

© 2025 The Author(s). Licensee IntechOpen. This chapter is distributed under the terms of the Creative Commons Attribution License (http://creativecommons.org/licenses/by/4.0), which permits unrestricted use, distribution, and reproduction in any medium, provided the original work is properly cited.

## References

[1] Weinstein L. Electromagnetic Waves. Moscow: Radio i svyaz'; 1988

[2] Hecht E. Optics, 5e. Uttar Pradesh: Pearson Education India; 2002

[3] Hudlicka M. Propagation of electromagnetic waves in periodic structures. [Thesis]. Prague: Czech Technical University;

[4] Weiglhofer WS, Lakhtakia A. Introduction to complex mediums for optics and electromagnetics. Washington, USA: SPIE press; 2003

[5] Engheta N, Ziolkowski RW. Metamaterials: Physics and Engineering Explorations. Canada: John Wiley & Sons; 2006

[6] Smith DR, Kroll N. Negative refractive index in left-handed materials. Physical Review Letters. 2000;**85**(14): 2933

[7] Veselago V. The electrodynamics of substances with simultaneously negative values of and. Uspekhi Fizicheskikh Nauk. 1967;**92**(3):517-526

[8] Smith DR, Pendry JB. Homogenization of metamaterials by field averaging. JOSA B. 2006;**23**(3): 391-403

[9] Cai W, Shalaev V. Optical Metamaterials. Fundamentals and Applications. New York, NY: Springer; 2010

[10] Schurig D, Mock JJ, Justice B, Cummer SA, Pendry JB, Starr AF, et al. Metamaterial electromagnetic cloak at microwave frequencies. Science. 2006; **314**(5801):977-980

[11] Leonhardt U. Optical conformal mapping. Science. 2006;**312**(5781): 1777-1780

[12] Smith DR, Liu R, Cui TJ. Metamaterials: Theory, Design, and Applications. US: Springer; 2010

[13] Pendry JB, Holden A, Stewart W, Youngs I. Extremely low frequency plasmons in metallic mesostructures. Physical Review Letters. 1996;**76**(25):4773

[14] Pendry JB, Holden AJ, Robbins DJ, Stewart WJ. Magnetism from conductors and enhanced nonlinear phenomena. IEEE Transactions on Microwave Theory and Techniques. 1999;**47**(11):2075-2084

[15] Shelby RA, Smith DR, Schultz S. Experimental verification of a negative index of refraction. Science. 2001; **292**(5514):77-79

[16] Ott HW. Electromagnetic Compatibility Engineering. Canada: John Wiley & Sons; 2011

[17] Paul CR, Scully RC, Steffka MA. Introduction to Electromagnetic Compatibility. Hoboken, USA: John Wiley & Sons; 2022

[18] Pal R. Electromagnetic, Mechanical, and Transport Properties of Composite Materials. Boca Raton FL, USA: CRC Press; 2015

[19] Kuila C, Maji A, Murmu NC, Kuila T, Srivastava SK. Recent advancements in carbonaceous nanomaterials for multifunctional broadband electromagnetic interference shielding and wearable devices. Carbon. 2023;**210**: 118075

[20] Geetha S, Satheesh Kumar K, Rao CR, Vijayan M, Trivedi D. EMI shielding: Methods and materials—A review. Journal of Applied Polymer Science. 2009;**112**(4):2073-2086

[21] Kondawar SB, Modak PR. Theory of EMI Shielding. Materials for Potential EMI Shielding Applications. UK: Elsevier; 2020. pp. 9-25

[22] Hong L, Lee DJ, Yoon HI, Han JS, Kim DJ. Influence of processing variables on scratch and shrinkage behaviors and translucency of dental zirconia. International Journal of Applied Ceramic Technology. 2020;**17**(1):354-364

[23] Singh SK, Sushmita K, Sharma D, Waidi YO, Bose S. Aerogels containing ionomers and microwave assisted growth of carbon nanostructures on carbon urchins for multifunctional electromagnetic interference shielding. Carbon. 2023;**209**:118036

[24] Acharya S, Ray J, Patro TU, Alegaonkar P, Datar S. Microwave absorption properties of reduced graphene oxide strontium hexaferrite/ poly (methyl methacrylate) composites. Nanotechnology. 2018;**29**(11):115605

[25] Zhu G, Giraldo Isaza L, Huang B, Dufresne A. Multifunctional nanocellulose/carbon nanotube composite aerogels for high-efficiency electromagnetic interference shielding. ACS Sustainable Chemistry & Engineering. 2022;**10**(7):2397-2408

[26] Mendoza-Sánchez B, Samperio-Niembro E, Dolotko O, Bergfeldt T, Kübel C, Knapp M, et al. Systematic study of the multiple variables involved in V2AlC acid-based etching processes, a key step in MXene synthesis. ACS Applied Materials & Interfaces. 2023;**15**(23):28332-28348

[27] Zhou Z-H, Liang Y, Huang H-D, Li L, Yang B, Li M-Z, et al. Structuring dense three-dimensional sheet-like skeleton networks in biomass-derived carbon aerogels for efficient electromagnetic interference shielding. Carbon. 2019;**152**:316-324

[28] Tripathi KC, Abbas SM, Sharma RB, Alegaonkar PS. Microwave absorbing properties of MWCNT/carbon black-PU Nano-composites. In: 2017 IEEE International Conference on Power, Control, Signals and Instrumentation Engineering (ICPCSI). Chennai: IEEE; 2017. pp. 489-496

[29] Yadav A, Tripathi KC, Baskey HB, Alegaonkar PS. Microwave scattering behaviour of carbon black/molybdenum di sulphide/cobalt composite for electromagnetic interference shielding application. Materials Chemistry and Physics. 2022;**279**:125766

[30] Alegaonkar AP, Alegaonkar PS. Nanocarbons: Preparation, assessments, and applications in structural engineering, spintronics, gas sensing, EMI shielding, and cloaking in X-band. In: Nanocarbon and Its Composites. UK: Elsevier; 2019. pp. 171-285

[31] Joshi A, Bajaj A, Singh R, Anand A, Alegaonkar PS, Datar S. Processing of graphene nanoribbon based hybrid composite for electromagnetic shielding. Composites Part B: Engineering. 2015; **69**:472-477

[32] Cui TJ, Liu R, Smith DR. Introduction to metamaterials. Metamaterials: Theory, Design, and Applications. 2010. pp. 1-19

[33] Holloway CL, Kuester EF, Gordon JA, O'Hara J, Booth J, Smith DR. An overview of the theory and applications of metasurfaces: The two-dimensional equivalents of metamaterials. IEEE Antennas and Propagation Magazine. 2012;**54**(2):10-35

[34] Zheludev NI. The road ahead for metamaterials. Science. 2010;**328**(5978): 582-583

[35] Hendry J. James Clerk Maxwell and the Theory of the Electromagnetic Field. (No Title). UK: A. Hilger; 1986

[36] Knott EF, Schaeffer JF, Tulley MT. Radar Cross Section. USA: SciTech Publishing; 2004

[37] Weston D. Electromagnetic Compatibility: Principles and Applications, Revised and Expanded. Boca Raton: CRC Press; 2017

[38] Landy NI, Sajuyigbe S, Mock JJ, Smith DR, Padilla WJ. Perfect metamaterial absorber. Physical Review Letters. 2008;**100**(20):207402

[39] Schade H. Schornsteinfeger US tech. mission to Europe. Tech. Rep. 90-45 AD-47746; 1945

[40] Du Toit LJ. The design of Jauman absorbers. IEEE Antennas and Propagation Magazine. 1994;**36**(6):17-25

[41] Lederer P. Modelling of practical Salisbury screen absorbers. IEE Colloquium on Low Profile Absorbers and Scatterers: IET. 1992:1/-/4

[42] Kaur R, Aul GD. Review on microwave absorbing material using different carbon composites. International Journal of Engineering Research & Technology (IJERT). 2014;**3**(5):160-167

[43] Deng G, Lv K, Sun H, Hong Y, Zhang X, Yin Z, et al. Wideband absorber based on conductive ink frequency selective surface with polarization insensitivity and wide-incident-angle stability. Nanomaterials and Nanotechnology. 2020;**10**: 1847980420935718

[44] Chen K, Luo X, Ding G, Zhao J, Feng Y, Jiang T. Broadband microwave metamaterial absorber with lumped resistor loading. EPJ Applied Metamaterials. 2019;**6**:1

[45] Priyanka MS, Alegaonkar PS, Baskey HB. Design and manufacturing of a Hexapattern frequency selective surface absorber for aerospace stealth application. ACS Applied Materials & Interfaces. 2023;**15**(30):37107-37115

[46] Alegaonkar PS, Baskey HB. Rheological investigations on frequency selective surface carbon composite microwave absorber. Materials Horizons. 2025;**12**(1):205-216

[47] Priyanka APS, Baskey HB. Design and architecting of a broadband bullet-diamond shaped integrated frequency selective meta-surface absorber for aero-stealth platform. Scientific Reports. 2024;**14**(1):23445

[48] Wang T, Han R, Tan G, Wei J, Qiao L, Li F. Reflection loss mechanism of single layer absorber for flake-shaped carbonyl-iron particle composite. Journal of Applied Physics. 2012;**112**(10): 104903-1-104903-6

[49] Abbas S, Chandra M, Verma A, Chatterjee R, Goel T. Complex permittivity and microwave absorption properties of a composite dielectric absorber. Composites Part A: Applied Science and Manufacturing. 2006; **37**(11):2148-2154

[50] Alitalo P, Tretyakov S. Electromagnetic cloaking with metamaterials. Materials Today. 2009; **12**(3):22-29

[51] Weiss M. Introduction to RF Linear Accelerators. Meyrin, Switzerland: CERN; 1994

[52] Li J, Pendry JB. Hiding under the carpet: A new strategy for cloaking. Physical Review Letters. 2008;**101**(20): 203901

[53] Hsu L, Lepetit T, Kanté B. Extremely thin dielectric metasurface for carpet

cloaking. arXiv preprint arXiv: 150308486. 2015

[54] Alitalo P. Microwave Transmission-Line Networks for Backward-Wave Media and Reduction of Scattering. [Thesis]. 2009

[55] Pattanaik B, Chauhan A. A study of stealth technology. Materials Today: Proceedings. 2023;**81**:543-546

[56] Pekker Á, Németh G, Botos Á, Tóháti HM, Borondics F, Osváth Z, et al. Cloaking by π-electrons in the infrared. Physica Status Solidi (b). 2016;**253**(12): 2457-2460

[57] Gharghi M, Gladden C, Zentgraf T, Liu Y, Yin X, Valentine J, et al. A carpet cloak for visible light. Nano Letters. 2011;**11**(7):2825-2828

[58] Alù A, Engheta N. Achieving transparency with plasmonic and metamaterial coatings. Physical review E —Statistical, nonlinear, and soft matter. Physics. 2005;**72**(1):016623

[59] Paul DD. Optical Metamaterials: Fundamentals and Applications. USA: American Institute of Physics; 2010

[60] Sarychev AK, Shalaev VM. Electrodynamics of Metamaterials. USA: World Scientific; 2007

[61] Shalaev VM, Cai W, Chettiar UK, Yuan H-K, Sarychev AK, Drachev VP, et al. Negative index of refraction in optical metamaterials. Optics Letters. 2005;**30**(24):3356-3358

[62] Ramakrishna SA. Physics of negative refractive index materials. Reports on Progress in Physics. 2005; **68**(2):449

[63] Veselago V, Braginsky L, Shklover V, Hafner C. Negative refractive index materials. Journal of Computational and Theoretical Nanoscience. 2006;**3**(2):189-218

[64] Smith DR, Pendry JB, Wiltshire MC. Metamaterials and negative refractive index. Science. 2004;**305**(5685):788-792

[65] Alu A, Engheta N. Plasmonic and metamaterial cloaking: Physical mechanisms and potentials. Journal of Optics A: Pure and Applied Optics. 2008; **10**(9):093002

[66] Pendry JB, Schurig D, Smith DR. Controlling electromagnetic fields. Science. 2006;**312**(5781):1780-1782

[67] Alù A, Engheta N. Multifrequency optical invisibility cloak with layered plasmonic shells. Physical Review Letters. 2008;**100**(11):113901

[68] Liu R, Ji C, Mock J, Chin J, Cui T, Smith D. Broadband ground-plane cloak. Science. 2009;**323**(5912):366-369

[69] Pradeep A, Baskey HB, Alegaonkar PS. Designing and development of ultrathin carpet cloak. In: Second International Conference on Current Trends in Physics and Photonics (ICCTPP 2024). Pune, india: SPIE; 2024. pp. 131-137

Section 3

# New Applications of Engineering Electromagnetic Field Technology

**Chapter 7**

# Perspective Chapter: Microwave Technological Processes

*Gennady Morozov, Yuri Sedelnikov, Vladimir Anfinogentov, Natalya Stakhova and Oleg Morozov*

## Abstract

The classification of microwave technological processes (MWTP) variants, based on the use of electromagnetic energy of the microwave range super and extremely high frequencies (SHF and EHF), allows us to distinguish at least two characteristic classes. The first is low-intensity (information) MWTP. This class strictly includes the processes characterized by the fact that the mass and phase states of the material remain unchanged during the processing, as do all its electro-physical parameters. For the specified MWTP, the phenomena of heat transfer and heat exchange with the environment do not play a significant role, or occur under conditions when the intensity of the electromagnetic field (EMF) and the thermos-physical parameters of the material remain practically constant. Thus, the analysis of electrodynamics and thermos-physical phenomena for them can be carried out independently. The second is high-intensity (thermal) MWTP. They are characterized by the fundamental necessity of joint consideration of electrodynamics and thermos-physical aspects. This requires a joint solution of the corresponding problems of EMF analysis, taking into account their changes due to heating and thermal fields influencing on changing distributions of EMF intensities. Despite this, we show conditions when it is possible to limit oneself to the analysis of only EMF. The chapter discusses both fundamental positions as practical realizations of low- and high-intensity MWTPs from the level of 35 years of R&D done in the Kazan microwave technologies science school.

**Keywords:** microwave technological processes, classification by structure, classification by intensity, joint consideration of electrodynamics and thermo-physical aspects, examples of information and thermal microwave technological processes

## 1. Introduction

It is impossible to imagine the modern world without various technical means based on the use of electromagnetic energy in the microwave range. Telecommunications, management of various processes, military affairs, transport, everyday life, and leisure—this is far from a complete list of areas in which the targeted impact of electromagnetic fields created by some technical means on other technical means is used. The rapid quantitative and qualitative growth of these means has also given rise to a number of negative consequences associated with the unintentional influence of EMF on equipment, materials, humans, and the environment.

The intentional impact of EMF created by artificial sources on technical means is the content of a vast field of science and technology—radio electronics, with its traditional division into more specific ones: radio communications, automation and computing technology, electronic warfare, etc. Unintentional impacts on technical means are considered within the framework of the problem of electromagnetic compatibility of technical means. Unintentional electromagnetic effects of technical means on biological objects constitute an independent field, which in recent years has been called electromagnetic ecology. The field related to the unintentional influence of EMF on materials and substances is usually not singled out as an independent field. The issues related to it are considered within the framework of design tasks, taking into account the influence of external factors.

Let us dwell on the identified groups of intentional effects of EMF on substances and biological objects. The field of activity operating with the use of the specified effects on substances, materials, and biological objects in order to change their properties or give them some new qualities has received the name *microwave technologies*.

The use of microwave technologies in industry, agriculture, and medicine is currently considered one of the most promising [1–3]. This is determined, at least, by two circumstances. Firstly, these are the features of heating materials by SHF EMF [4, 5]:

- heating is caused by the release of heat in the volume of the processed material under the influence of EMF;

- there are wide possibilities for controlling heating, both by changing the oscillation frequency and by controlling the excitation of the working chambers;

- the possibility of selective action on heterogeneous materials, namely more intense heating of parts with high specific absorption;

- heating is practically inertia less, since the release of heat stops immediately with the cessation of electromagnetic action.

Secondly, the impact of EMF leads in some cases to results that are not equivalent to heating an object to the same temperature by traditional means [6–8]. This is especially true for biological objects. The impact of EMF is accompanied by a wide range of effects depending on the frequency range, intensity, duration of exposure, and modulation parameters. At higher EMF intensity, as a rule, suppressive effects are observed, up to the death of biological objects. At moderate and low intensities, in some cases, stimulating effects are observed, leading to accelerated development, increased immune functions of organisms, etc. In particular, this applies to the impact of EMF in the millimeter wavelength range on a wide range of biological objects—from plants and microorganisms to higher animals and humans.

The listed groups have an important distinctive feature. Any effects associated with the heating of objects by EMF can be quantitatively described on the basis of known laws of physics. Despite the extreme complexity of the existing computational procedures, methods of computational electrodynamics and thermodynamics allow, in principle, the calculation of the mechanisms of microwave heating with any accuracy. With regard to biological effects, the situation is fundamentally different—most of them cannot be explained at present within the framework of existing knowledge.

Nevertheless, the accumulated experimental data provide sufficient grounds for the practical use of the identified effects of SHF and EHF EMF on various biological objects.

The main areas of using SHF and EHF EMF should be classified according to features associated with the duration and intensity of exposure. The first group should include various options for drying materials, heating, defrosting processes, roasting materials, cooking, etc. These processes are characterized by a high intensity of EMF and duration of exposure, leading to significant heating of the processed material and heat exchange with the environment. A number of them are also characterized by a change in the phase state of the processed substances, accompanied by a significant change in the electrical and thermal parameters of the material. Finally, in the processes of microwave drying of various materials, not only changes in the electro-physical parameters occur but also in the mass of the processed material as moisture is lost. To maintain the conditions of uniform processing during these processes, it is necessary to create nonuniform and, moreover, time-varying EMFs.

The listed cases determine the class of applications called *high-intensity (thermal) microwave technologies*. The analysis and optimization of processes of this class are characterized by the need for a joint consideration of electrodynamic and thermal physical aspects, including heat and mass transfer phenomena. Despite the exceptional complexity of the mathematical description, work in this direction is being intensively carried out, and by now, a number of fundamental results have been obtained.

Another large group is made up of applications that make up the class of *low-intensity (information) microwave technologies*. These processes are characterized by very low levels of intensity of EMF that do not lead to significant heating of the processed materials. Examples of this kind can be various options for using the stimulating effects of the action of extremely high frequency EMF on biological objects. In addition, this class can include processes of short-term SHF processing of EMF of moderate and even, in some cases, high intensity. Common to the processes of this class is that the processes of heat transfer and heat exchange with the environment no longer play a significant role here, and the electro-physical parameters during the processing remain virtually constant. Thus, when analyzing processes of this class, there is no need to consider electrodynamic and thermos-physical processes together.

Another important consequence is that to implement a low-intensity MWTP, it is necessary to create EMFs that are homogeneous in the volume of the material. This circumstance, on the one hand, significantly simplifies the analysis and subsequent optimization of processes in the working chambers of installations compared to similar tasks for high-intensity technologies. On the other hand, the phenomenon of heat transfer under long-term exposure to EMF promotes equalization of thermal field distributions, which somewhat simplifies the designer's task. Under short-term exposures, the presence of areas with insufficient or excessive field intensity means that in the specified parts of the material, the electromagnetic action will either not produce the desired effect or will lead to the deterioration of the properties. Thus, low-intensity microwave technologies are characterized by rather strict requirements for the degree of homogeneity of the electromagnetic fields used.

By now, a number of technological processes from the class of high-intensity microwave technologies have been implemented in industrial installations produced in series. Low-intensity microwave technologies, despite the promising results of scientific research, have not yet received due distribution. Possible reasons include insufficiently developed methodology for creating microwave installations, as well as subjective factors. The purpose of this chapter is to discuss the development paths aimed at creating effective technical paths for implementing low-intensity microwave technological processes as complex ones in agriculture and industry.

## 2. Microwave technological processes

### 2.1 Classification of microwave technological processes

The field of applications in which the effective use of SHF and EHF electromagnetic energy is possible is very wide. The intensity of the EMF, the duration of the treatment, and its result—changes in the properties of the material being treated—also vary widely. The following characteristic groups can be distinguished among the described applications.

1. The intensity of the EMF is very low, and the treatment does not lead to any significant heating of the material. This group mainly includes various options for the effect of the EMF of the SHF range on biological objects with the purpose of their stimulation or suppression: pre-sowing treatment of seeds, stimulation of biochemical processes, therapy, etc.

2. The treatment is carried out by the EMF of the microwave range for a relatively short period. In these cases, the material is heated; however, due to the short duration of the process, the effects of heat transfer and, especially, heat exchange with the environment do not play a decisive role. As a result, the spatial distribution of EMF intensities and temperature fields are close to each other, which allows us to limit ourselves to EMF analysis, that is, solving electrodynamic problems. An example is the stimulation of seeds by microwave EMF, some problems of material disinfection, etc.

3. The treatment is carried out by microwave EMF of relatively low intensity over a long period. A characteristic feature of this group of effects is that heat transfer processes play a significant role, because of which the spatial dependences of temperature fields and EMF intensities differ significantly. This requires solving both the corresponding electrodynamic and thermos-physical problems. The concept of "relatively low intensity" of EMF in this case means that as a result of heating, there is no significant change in the electrical parameters of the material, nor, moreover, a change in its mass, composition, and phase state. Examples of such processes include most cases of food disinfection.

4. The treatment is carried out by high-intensity EMF over a long period. Because of heating, the electro-physical parameters of the material change significantly, which leads to a significant change in the EMF pattern during the treatment. Typical examples of this group are various options for using EMF energy for the polymerization of plastics, vulcanization of rubber, etc.

5. High-intensity and long-term EMF treatment, leading to a change in the phase state of the substance. The peculiarity of this group of applications is that when the phase state of most materials changes, a more significant and abrupt change in the electro-physical parameters occurs than in the previous case—both

specific conductivity and permittivity and thermal physical parameters. Examples—defrosting of products, melting of various materials, etc.

6. Long-term high-intensity EMF treatment, leading to a change in the mass and electro-physical parameters of the substance. A typical example is the microwave drying of various materials. The most important feature of such processes is the fundamental need to take into account the above changes, both in the analysis of EMF and thermos-physical processes.

The classification of various options for implementing microwave technological processes (MWTP) based on the use of electromagnetic energy of the SHF and EHF ranges allows us to distinguish at least two characteristic classes.

The first of them is low-intensity MWTP. This class strictly includes the processes according to p.1 and, with some conventionality, according to pp. 2–3. They are characterized by the fact that the mass and phase state of the material remain unchanged during the processing, as do all its electro-physical parameters. For the specified MWTP, the phenomena of heat transfer and heat exchange with the environment do not play a significant role or occur under conditions when the intensity of the EMF and the thermos-physical parameters of the material remain practically constant. Thus, the analysis of electrodynamic and thermos-physical phenomena for them can be carried out independently: the analysis of EMF without taking into account heating factors and the analysis of thermos-physical processes for given EMF without taking into account their changes due to heating effects. Moreover, for MWTP, according to pp. 1–2, it is possible to limit oneself to the analysis of only EMF.

High-intensity MWTP includes various cases according to pp. 4–6. They are characterized by the fundamental need for a joint consideration of electrodynamic and thermos-physical aspects. This requires a joint solution to the corresponding problems of analyzing EMF, taking into account their changes due to heating and thermal fields, the changing mass of the material, its thermos-physical parameters, and heating conditions with changing distributions of EMF intensities.

The share of electromagnetic energy used in them can divide MWTP. The classification of MWTP by this feature is shown in **Figure 1**.

The MWTP for which the given properties are imparted to the processed object by exposing it only to SHF and EHF electromagnetic fields will be called unitary, and the microwave process technology for which the given properties are imparted to the processed object by exposing it to both microwave and other physical factors will be called combined. In turn, combined MWTP can be divided into parallel and sequential depending on the sequence of exposure of the processed object to electromagnetic fields

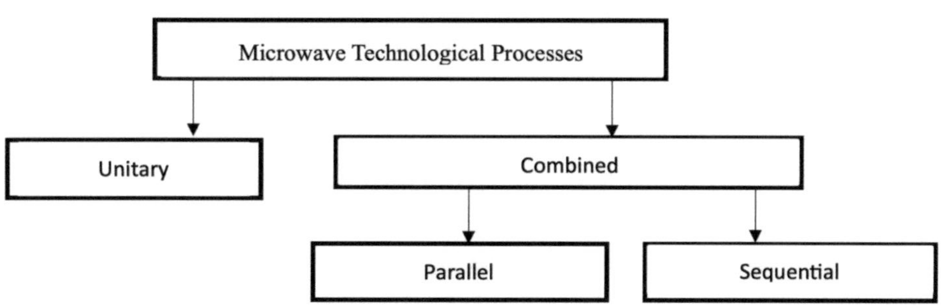


**Figure 1.**
*Classification of microwave technological processes.*

and other physical factors. When they are applied simultaneously, the processes will be called parallel, and when applied sequentially, the processes will be called sequential.

Let us consider the essence of the specified MWTP in more detail.

## 2.2 Types of microwave technological processes

### *2.2.1 Unitary microwave technological processes*

It is typical for a unitary MWTP that the process of imparting specified properties to the object being processed is achieved by exposing it only to SHF and EHF EMF. A generalized diagram of such an impact is shown in **Figure 2**.

The initial object (material, product, etc.) is characterized by a certain vector of indicators $\mathbf{X_{in}}$, which include both qualitative (humidity, biological properties, temperature, etc.) and quantitative (mass, volume) indicators. A vector of indicators $\mathbf{V}$, which generally describes the frequency, intensity, polarization, modulation law, duration of exposure, etc., characterizes the acting electromagnetic field. Because of exposure to electromagnetic fields, the object being processed acquires a new set of properties, described by a vector of indicators $\mathbf{X_{out}}$ (the composition and nomenclature of the $\mathbf{X_{out}}$ constituents do not necessarily correspond to the components $\mathbf{X_{in}}$ and may differ from the latter both in composition and in physical meaning).

The result of microwave processing is determined by the operator transformation $\mathfrak{R}_{un}$ as

$$\mathbf{X_{out}} = \mathfrak{R}_{un}(\mathbf{X_{in}}, \mathbf{V}) \tag{1}$$

and depends on the properties of the initial object and the parameters characterizing the electromagnetic effect.

The quality of a unitary MWTP can be estimated by comparison $\mathbf{X_{out}}$ with a given set of output properties $\mathbf{X^*_{out}}$, and the necessary effects are defined as the values of the action vector that ensure the minimum difference between the output properties and the specified ones,

$$\left\|\mathbf{X_{out}} - \mathbf{X^*_{out}}\right\| \to \min_{\mathbf{V}\in\Omega_v} \tag{2}$$

under the condition that the control actions $\mathbf{V}$ belong to a certain set of permissible values $\Omega_v$.

In the general case, it should be considered that the quality of the MWTP can be described by a certain technical and economic vector criterion $Q_\Sigma(\mathbf{X_{in}}, \mathbf{X_{out}}, \mathbf{V})$ taking

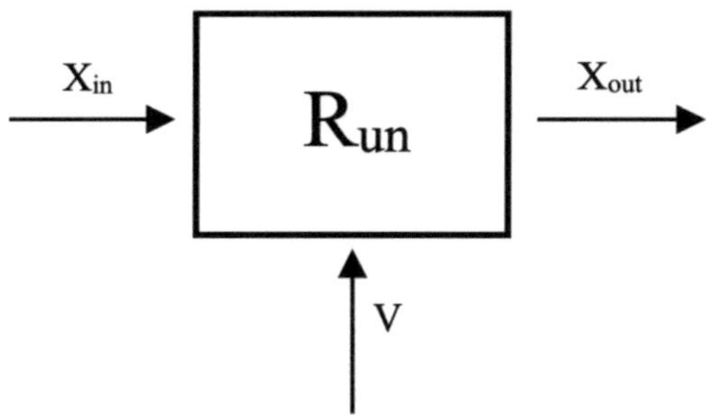


**Figure 2.**
*A generalized diagram of unitary MWTP.*

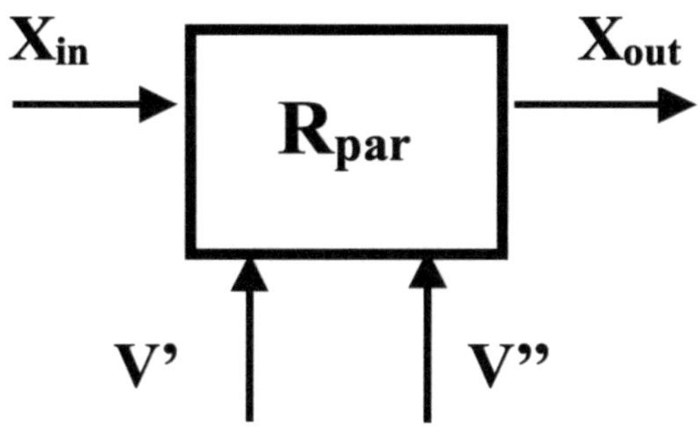


**Figure 3.**
*A generalized diagram of combined MWTP of parallel type.*

into account various requirements for the initial and output products, as well as the parameters of the control actions. In the general case, the control parameters of such a process should be selected based on the conditions of the extremum of the specified criterion under permissible effects

$$Q_{\Sigma}(\mathbf{X_{in}}, \mathbf{X_{out}}, \mathbf{V}) \rightarrow \max_{\mathbf{V} \in \Omega_v}(\min) \tag{3}$$

The actual implementation of the process, including the selection of its parameters according to the criteria (2)–(3), is a nontrivial task and will be considered below.

### *2.2.2 Combined microwave technological processes*

For a number of applications, in particular drying of various materials, the combined use of the effects of EMF and other physical factors is typical, for example, convective heating, infrared (IR) heating, low pressure, etc. It is the combined use of several physical effects in a number of technological processes that allows for the most efficient processing, thereby ensuring the best technical and economic indicators of it [1–3]. Let us consider similar processes schematically.

The initial product, as before, will be characterized by a vector of indicators $\mathbf{X_{in}}$, the final product—$\mathbf{X_{out}}$. Physical effects in the processing process are determined by sets of parameters corresponding to the effects of EMF and other physical factors—$\mathbf{V}'$ and $\mathbf{V}''$. According to the available literature data [2], two types of combined technologies can be distinguished, in the first of which the impact of EMF and other physical factors is carried out simultaneously, in the second—in turn. We will call such processes, respectively, combined MWTP of parallel and sequential types.

*Combined MWTP of parallel type*. For the process of parallel type, the implementation scheme is shown in **Figure 3**.

The result of combined processing $\mathbf{X_{out}}$is some operator transformation $\mathfrak{R}_{\mathbf{par}}$ of the initial vector of indicators $\mathbf{X_{in}}$

$$\mathbf{X_{out}} = \mathfrak{R}_{\mathbf{par}}(\mathbf{V}', \mathbf{V}'', \mathbf{X_{in}}) \tag{4}$$

The task of implementing such a process consists of the joint selection of control actions, characterized by vectors $\mathbf{V}'$ and $\mathbf{V}''$, providing, in the general case, the best technical and economic indicators of the process as a whole:

$$Q_\Sigma(\mathbf{X_{in}}, \mathbf{X_{out}}, \mathbf{V}', \mathbf{V}'') \rightarrow \max_{\mathbf{V}' \in \Omega_{v1}, \mathbf{V}'' \in \Omega_{v2}} (\min) \tag{5}$$

The peculiarity of this problem also consists of the fact that for the developer of the process, it is necessary to select the best combination of the effects of microwave electromagnetic energy and other types of effects in order to ensure the best technical and economic indicators of the process (5) as a whole.

To date, problems in such a formulation have not been set for sufficiently wide classes of technological processes. Moreover, for many processes of practical interest, the corresponding approaches have not been sufficiently developed. Therefore, for combined MWTP of this type, it must be assumed that the degree of reasonable combination of the effects of electromagnetic energy and other factors is determined empirically by specialists in a specific subject area. This allows the designer of the MWTP to consider the values of the control parameters known $\mathbf{V}'' = \mathbf{V}''^*$, as well as the values of the indicators characterizing the electromagnetic effects determined within certain limits $\mathbf{V}' \in \Omega_{v1}$. With these assumptions, the problem of selecting the best indicators of electromagnetic effects takes the form

$$\mathbf{X_{out}} = \mathfrak{R}_{\mathbf{par}}(\mathbf{V}', \mathbf{V}'', \mathbf{X_{in}})\Big|_{\mathbf{V}'' = \mathbf{V}''^*} = \mathfrak{R}_{\mathbf{par}}(\mathbf{V}', \mathbf{X_{in}}). \tag{6}$$

*Combined MWTP of a sequential type*. In combined sequential-type microwave thermal processing plants, the EMF with parameters and other physical factors determined by the indicators are alternately exposed. A similar process can be schematically shown in **Figure 4**.

As in the case of combined parallel type microwave thermal processing plants, the issue of the optimal choice of the number of stages and the ratio of the effects of the EMF and other physical factors at the corresponding stages has not yet been sufficiently studied. Nevertheless, specialists in the subject area (e.g., grain drying) have empirically determined some reasonable values of the specified indicators. Thus, for the designer of the microwave thermal processing plant, with a certain degree of conventionality, it is necessary to consider the approximate values $\mathbf{X_{in(N)}}, \mathbf{X_{out(N)}}$ of the indicators for each of the process steps $S = S_1 \dots S_N$ in which the EMF is exposed to are known.

Thus, the task of choosing microwave processing modes is to ensure that the best technical and economic indicators are provided at each of the stages corresponding to it.

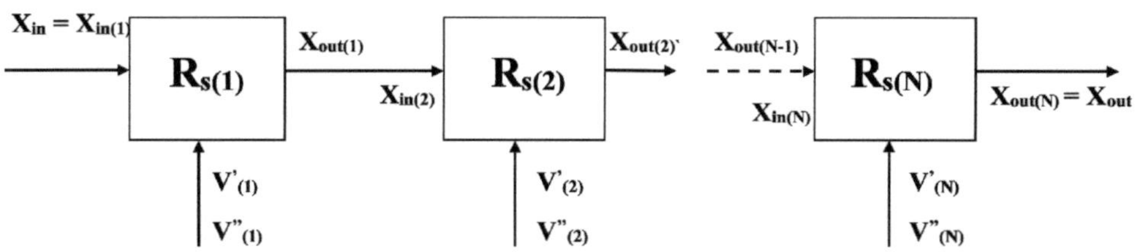


**Figure 4.**
*A generalized scheme of the combined MWTP of a sequential type: $\mathbf{X}_{out(N-1)} = \mathbf{X}_{in(N)}$, where $\mathbf{X}_{in(N)}$—vector of indicators of the input properties of the object before stage N; $\mathfrak{R}_{s(N)}$—operator of the converting of a combined sequential type at the stage N; $\mathbf{X}_{out(N-1)}$—vector of indicators of the output properties of the object after the stage N-1; $\mathbf{V}'_{(N)}$—vector of EMF effects at stage N, $\mathbf{V}''_{(N)}$—vector of indicators of additional physical impact at stage N.*

$$Q_{\Sigma}^{(S)}\left(\mathbf{V}'_{S}, \mathbf{X}_{\mathbf{вх}}^{(\mathrm{S})}, \mathbf{X}_{\mathbf{вых}}^{(\mathrm{S})}\right) \to \max_{\mathbf{V}'_{s} \in \Omega_{vs}} (\min) \tag{7}$$

### *2.2.3 Generalized scheme of the MWTP*

As follows from the previous sections, the essence of the MWTP (1) can be described as an operator transformation $\mathfrak{R}$ of the initial product with properties $\mathbf{X_{in}}$ into a product with a set of properties $\mathbf{X_{out}}$, where $\mathbf{V}$ is a set of parameters of the control actions of the EMF during the process (or its S-th stage).

Let us clarify the introduced category of control parameters of the process **V**, as technological factors. Let us distinguish three groups of parameters.

The first **V1** group describes the parameters that determine the course of the MWTP: the intensity of the EMF, polarization, time dependence, including the modulation law, etc. In this group of parameters, it is advisable to divide it into two subgroups.

The first subgroup **V1E** refers to the elementary volume of the processed material and determines for it the values of the direction of the electric field, polarization, modulation law, and the change in field intensity over time. By elementary volume, we shall mean a small spatial region within which spatial changes in the EMF parameters can be neglected, considering their values to be constant. Thus, the group of parameters **V1E** determines the necessary parameters of EMF action to implement the required transformations of the properties of the original product.

The second subgroup—**V1V**—refers to the entire volume of the processed material and is related to the spatial dependencies and time changes in the EMF parameters in the entire volume of the processed material. Such an introduction of parameters is advisable for two reasons. Firstly, it allows us to independently consider the issues of optimizing the EMF parameters of both subgroups **V1E** and **V1V** required to implement the transformation $\mathbf{X_{in}} \to \mathbf{X_{out}}$. Indeed, **V1E** does not include indicators of the spatial distribution of fields in the processed volume, and **V1V** can operate with relative values of the spatial distributions of monochromatic fields without taking into account the presence and parameters of modulation.

The second group of indicators, **V2**, are the quantities that determine the process of creating the required EMF (with a given time dependence, intensity, and spatial distribution). Depending on specific conditions, these quantities may include the parameters of generators and modulators, as well as the dimensions of the working chamber, exciter parameters, etc.

The third group of indicators, **V3**, describes the technical parameters of a specific installation. When implementing a MWTP by a specific microwave technological complex (MWTC), the required transformation is ultimately determined by the parameters **V3**, since

$$\mathbf{V}_1 = \mathbf{V_1}(\mathbf{V}_2), \mathbf{V}_2 = \mathbf{V}_2(\mathbf{V}_3), \mathbf{V_1} = \mathbf{V}_1(\mathbf{V}_2(\mathbf{V}_3)). \tag{8}$$

However, given the extreme complexity of the MWTC optimization processes, it is advisable to consider the problem in stages, introducing various indicators of control actions at different hierarchical levels:

$$\mathbf{V}_1\mathbf{E} \to \mathbf{V}_1\mathbf{V} \to \mathbf{V}_2 \to \mathbf{V}_3 \tag{9}$$

The main attention will be paid to the issues of determining the parameters of group **V1**.

The task of the best organization of the generalized MWTP is the optimal choice of control actions in accordance with the conditions for achieving the best technical and economic indicators (3). In this case, a fundamentally important issue that requires careful consideration is the selection and formalization of requirements that define the concept of technical and economic indicators. The following section is devoted to examining approaches to constructing these indicators (i.e., criteria).

### 2.3 Criteria for assessing the efficiency of a MWTC

The task of MWTC construction in its most general form is to select conditions under which the overall quality is the best. At present, in the stated general formulation, such problems can only be solved with significant simplifications and assumptions. The reason for this is not only and not so much the multi-criteria nature of the problem. In our opinion, the reasons are of a deeper nature. Let us list these reasons.

Firstly, in a number of cases, MWTCs allow the implementation of useful effects that are not achievable with traditional technologies. Secondly, some important MWTC indicators are practically not amenable to formalization. These include various social aspects of microwave technologies (MWT), such as the possibility of reducing environmental pollution by reducing the use of chemicals, increasing the environmental friendliness of products, increasing the prestige of work using MWT, etc. The complexity of the processes, as well as the listed circumstances, do not allow for the full application of economic criteria based on a comparison of the positive effect of using a new technological process with the costs of its implementation. At the same time, the cost factor is extremely important: more the wider use of the MWT is constrained by the relatively high cost of the equipment. Therefore, it is very important to implement a specific MWT in such a way that the total costs $C_{\Sigma}$ are minimal for fixed volume and quality of processing provided that

$$\|\mathbf{X}_{\mathbf{out}} - \overset{*}{\mathbf{X}}_{\mathbf{out}}\| \leq \varepsilon \tag{10}$$

where $\overset{*}{\mathbf{X}}_{\mathbf{out}}$ a set of parameters corresponding to the best processing conditions determined without taking into account cost factors. However, the problem of optimizing the construction of the MWTC in the form (9) cannot be brought to the final calculations in the general case. It is necessary to find a reasonable way to introduce simplifying provisions. The latter can be constructed on the following basis.

Regardless of the specific purpose, volume of output, etc., the total costs $C_{\Sigma}$ for the creation and use of a certain MWT are a monotonic function of the costs of electromagnetic energy of microwaves per unit of output. Therefore, a reasonable simplifying position is the condition of implementing the microwave thermal process in such a way that with a fixed quality of conversion, the minimum costs of specific microwave electromagnetic energy $W_S$ are ensured

$$W_S = \frac{W_T}{N} \to \min, \tag{11}$$

where $W_T$ is the total energy cost for the production of a conventional batch of products from $N$ items.

Here, however, it is necessary to make a very significant reservation. As is known, for the implementation of the microwave thermal process, a number of frequency ranges have been allocated, corresponding to the decimeter, centimeter, and millimeter ranges (435, 985, 2450, 22,500 MHz, etc.). At the same time, the cost of producing 1 J of energy turns out to be different for different frequency ranges. At the same time, the technical efficiency of a specific technological process can strongly depend on the frequency range. Therefore, the criterion of minimum specific energy consumption should be modified in order to take into account the listed circumstances: use as a criterion the value (11), multiplied by weighting coefficients $a_f$ and $a_m$ to take into account, respectively, the difference in the cost of implementing generators in different frequency ranges and with different types and parameters of modulation

$$W_R = W_S a_f a_m \tag{12}$$

Hereinafter, the value $W_R$ will be called the reduced specific energy consumption of electromagnetic energy. The physical meaning of the value (12) is as follows: the reduced energy consumption is the value of the total energy consumption that ensures the process of microwave processing of a unit of production, taking into account the amendments that the change in the cost of implementation depending on the frequency range, type and parameters of modulation.

## 3. Problems of choosing the information processing mode

When creating a MWTC of both unitary and combined types, the designer uses the data on the parameters of the electromagnetic effect—**V1**—as the initial data. As shown in Section 2, it is advisable to divide these parameters into two groups—**V1E** and **V1V**. The first of them refers to idealized processing conditions in a hypothetically homogeneous EMF (or, equivalently, EMF for an elementary volume of the processed material). The second group characterizes the processing parameters in real conditions, in particular, the heterogeneity of the EMF.

The division into these groups is due to a number of reasons and also has the peculiarity that the parameters **V1E** are determined at the stage of predesign studies, and the parameters **V1V**—at the stage of R&D to create a specific MWTP and the MWTC implementing it. It should be emphasized that, in most cases, the predesign research stage is carried out by specialists professionally oriented in the field of applications (biologists, chemists, etc.), while the R&D stage is carried out by radio engineers. This leads to the fact that the recommended processing conditions (specified by the **V1E** parameters), at best, correspond to the conditions of optimal processing quality, that is, are determined without taking into account the specific features of subsequent implementation.

### 3.1 Main decision

The energy consumption factor and, especially, the reduced microwave energy consumption often remain out of sight. In accordance with the approach described in Section 2, a somewhat different task should be solved at the predesign research stage. Namely, such conditions of electromagnetic processing **V1EO** (optimal) are determined, which correspond to the minimum of the reduced costs of electromagnetic energy

$$W_S \rightarrow \min \tag{13}$$

with acceptable quality of processing $Q_A$

$$Q(\mathbf{V_1EO}) \geq Q_A \tag{14}$$

or under the condition

$$\left| Q\left(\overset{*}{\mathbf{V_1E}}\right) - Q(\mathbf{V_1EO}) \right| \leq \varepsilon \tag{15}$$

where $Q\left(\overset{*}{\mathbf{V_1E}}\right)$ is the best value of the quality indicator of processing, determined without taking into account the factor of energy costs.

Finding the specified parameters **V1EO**, optimal from the point of view of the ratio "quality - energy costs," will be called determining the microwave processing mode. Examples and possibilities of effectively finding such modes, corresponding to low-intensity MWTP, are considered for sowing properties of seeds, as an example of low-intensity "information" technology.

The sowing properties of seeds are determined by two characteristics: germination energy and germination. Germination energy determines the vitality of seeds and their ability to germinate. Germination characterizes the ability of germinated seeds to further develop (form a full-fledged plant). Of course, the yield depends on the sowing properties. In addition, for many crops, for example, cereals, seeds with a germination rate of (90–95)% make up only (40–45)% of the total volume. Thus, the seeding rate is constantly overestimated, which leads to high economic costs. Therefore, stimulation of seeds in order to improve their sowing properties seems to be very relevant.

For seeds of any crop, the relationship between the quality of microwave treatment and the energy expended has the form of a dependence shown in **Figure 5**. The effect of improving seed properties appears approximately six hours after treatment and increases with longer storage, reaching a maximum after 20–30 days. Then, the effect stabilizes and lasts for up to two months. Treatment is carried out with parameters corresponding to heating of seeds to temperatures of about 45–50°C, that is, at a high intensity of EMF (about 0.1–1 W/cm$^2$).

When treating seeds with EMF of the extremely high frequencies range, the stimulation effect is manifested to at least no lesser extent. Germination and

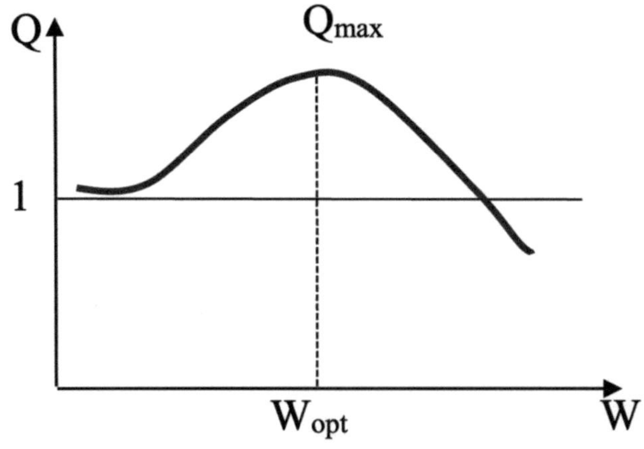


**Figure 5.**
*Dependence of the quality of microwave processing vs. the energy expended in the microwave range.*

germination energy increase. In addition, with extremely high frequencies treatment, the resistance of seedlings to diseases affecting plants at an early stage of development increases to a greater extent. A possible explanation for this fact is that, as a result of microwave treatment, seeds germinate faster than pathogens have time to develop.

Regardless of the type of crop, extremely high frequencies seed treatment has a number of common patterns:

- the effect is manifested when treated with EMF with a frequency of about 42 and 53 GHz;
- there is a threshold level of field intensity, above which the stimulation effect is observed. The value of this threshold is no more than 10 $\mu W/cm^2$;
- there is a minimum exposure time required to achieve a useful result. The specified time is practically independent of the field intensity exceeding the threshold level. A further increase in the treatment time in some cases can lead to the opposite effect—deterioration of sowing properties;
- the effect of exposure to extremely high frequencies EMF is manifested almost immediately (i.e., long-term seed curing is not required) and persists for a fairly long time (at least a month).

There is a (though limited) number of studies that demonstrate the effectiveness of using multifrequency treatment for seed stimulation purposes. The combined use of centimeter and millimeter fields were studied. According to these studies, combined multifrequency treatment of seeds can lead to additional improvement of sowing properties.

### 3.2 Practical examples of information processing. Pre-sowing seed treatment

For the first time, the effect of microwave electromagnetic energy was tested for scarification purposes, that is, for the destruction of the hard seed coat of some crops [9–12]. The next step was heating the seeds with EMF, which repeated in its purposes a long-known agricultural technique—heating the seeds at a temperature of about 40–45°C. It was soon discovered that the effect of stimulating the sowing properties when exposed to microwave EMF is more pronounced than when heated by traditional means to the same temperatures. Numerous studies conducted in this area have made it possible to reliably assert that as a result of exposure to decimeter wave (SHF) EMF, at certain field intensities and durations of exposure, the sowing properties of seeds improve and the sprouts develop faster, which ultimately leads to an increase in crop yields. Since the 1970s, the effects of exposure to EMF in the short-wave millimeter wave range (EHF) (∼40–60 GHz) have attracted the attention of specialists. It was found that under certain conditions, pronounced stimulating effects are observed, including improved sowing properties of seeds of various crops. It is important to note that these effects are the result of exposure to EMF of very low intensity (less than 10 $\mu W/cm^2$), have a frequency dependence of a resonant nature with a "quality factor" of about $10^2$. Attempts have also been made to identify more subtle effects. Thus, for therapeutic applications, it is advisable to use frequency-modulated (FM) oscillations. When treating seeds with EMF in the SHF range, it is preferable to use circular polarization fields and of a certain direction of rotation.

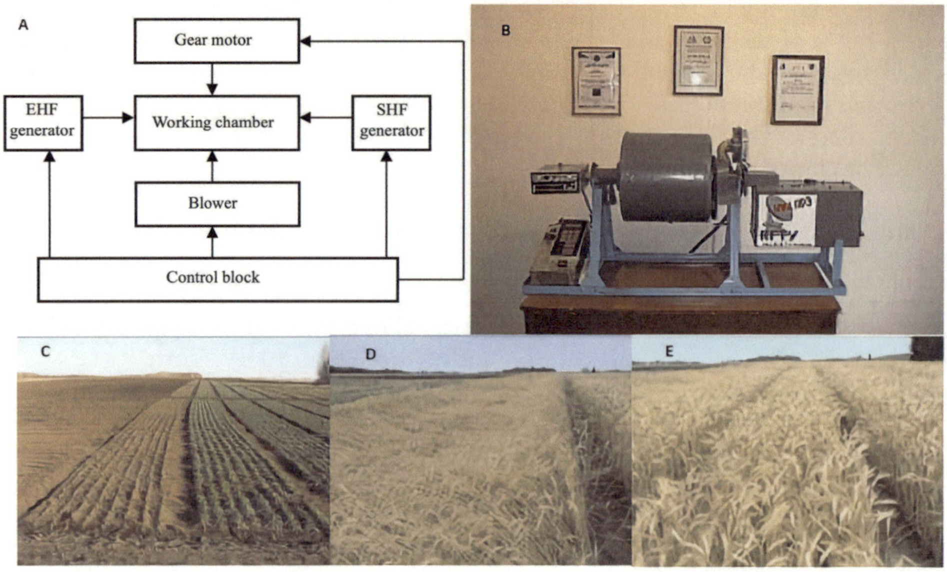


**Figure 6.**
*Dual-range microwave unit for pre-sowing seed treatment "SHYTYM": a generalized scheme (A); microwave unit photo (B); sprouted seedlings of winter wheat Kazan-560—weak in the control band without treatment on the left and strong in the band with treatment on the right (C); ears of winter wheat Kazan-560—empty, lodging in the control band without treatment (D) and filled, stable in the band with treatment (E).*

The dual-range microwave unit (**Figure 6**) for pre-sowing seed treatment "SHYTYM" operates in two ranges: SHF and EHF.

SHF range: frequency 2450 MHz, productivity 24 kg/hour; power consumption no more than 0.3 kW/hour.

EHF range: frequencies (30 ... 80) GHz, productivity 10 kg/hour, power consumption no more than 0.05 kW/hour.

Pre-sowing treatment of agricultural crop seeds with electromagnetic fields of the SHF and EHF ranges improves their sowing qualities, germination energy, stability to lodging, resistance to infectious diseases, which leads to increased yields.

## 4. Problems of electromagnetic and thermal field generation

### 4.1 Main types of problems

When creating real MWTCs that process materials using SHF and EHF electromagnetic energy, one of the fundamentally important issues is the generation of an EMF with the required properties in the working chamber. Above, in Section 3, the issues of selecting the MWTP mode were considered. When creating an MWTC, these data (in the notation of Section 2—parameters **V1E**) are specified values. Depending on the specific problem, these include duration of exposure, field intensity, heating temperature, etc. It is believed that if, during implementation, these parameters exactly correspond to the specified values for all points in the volume of the material being processed, the quality of the MWTP will be the best. At the same time, with the exception of the oscillation frequency and processing time, all parameters of the electromagnetic effect significantly depend on the size and shape of the working chamber, its load, the electro-physical parameters of the material, and the properties of the devices that excite the EMF.

Due to a number of fundamental reasons, it is impossible to ensure uniform processing conditions throughout the entire volume of material. As a result, the parameters affecting the quality of EMF processing at different points in the working chamber will have values that differ from the given parameters **V1E**. These differences, which determine deviations from the specified mode, are characterized by some parameters **V1V**. This Section provides materials that allow us to establish a connection between the processing quality indicators for the entire volume of material being processed with the parameters of the processing mode **V1E** and the indicators **V1E**. We also consider ways of forming an EMF with the required properties on the example of the probabilistic approach.

Let us first specify the main types of tasks corresponding to the formation of an EMF for the MWTP. Depending on its goals and the properties of the material, two main types can be distinguished.

Let us classify all cases in which the requirements for the MWTP parameters can be formulated as requirements for the spatial distribution of the EMF in the volume of the material to the first of them, E-class [3]. In most cases, the requirements for the spatial distribution of the field $W\left(\vec{r}\right) = \sigma\left|\vec{E}\left(\vec{r}\right)\right|^2$ are determined by the values of the given specified absorbed power $WG\left(\vec{r}\right)$:

$$W\left(\vec{r}\right) = \sigma\left|\vec{E}\left(\vec{r}\right)\right|^2 \rightarrow WG\left(\vec{r}\right) \tag{16}$$

There may also be situations when the required EMF must have a certain vector structure:

$$\vec{E}\left(\vec{r}\right) \rightarrow \vec{E}_G\left(\vec{r}\right) \tag{17}$$

For most low-intensity MWT, homogeneous EMFs are specified.

T-class problems consist of forming such EMFs that correspond to thermal fields of the required spatial configuration:

$$T\left(\vec{r}\right) \rightarrow T_G\left(\vec{r}\right) \tag{18}$$

We divide the T-class problems into two types: T1 and T2.

The first type T1 corresponds to the excitation of such an EMF that minimizes the deviation of the thermal field in the processed volume in the steady-state mode from the specified temperature distribution.

$$t_{ss} \rightarrow \min_{\vec{E}\left(t, \vec{r}\right)} \tag{19}$$

The second type T2 consists of forming such an EMF with a time-varying intensity that minimizes the duration of the transition to the steady-state thermal field under conditions of forming a specified temperature distribution and limited input power $P_L$:

$$\left\| \lim_{t\to\infty} T\left(\vec{r}\right) - TG\left(\vec{r}\right) \right\| \le a_1, \ \max_t P_L(t) \le a_2 \tag{20}$$

where $a_1$ and $a_2$ are specified values.

Further in this Section, only E-class problems are considered, as they correspond to the category of "low-intensity microwave technologies".

### 4.2 Types of microwave processing and basic requirements for EMF parameters

Analysis of the current state of MWT [13–33] shows that there are various options for microwave processing, which differ significantly in the nature of the requirements for the level and duration of electromagnetic effects. Despite the significant diversity of specific applications, there are only two main types of low-intensity MWTP.

The first of these includes various cases of short-term $T_{PR}$ heating of the material by microwave EMF. In this case, it is advisable to use the value of the specific absorbed energy as a criterion:

$$W = \int_0^{T_{PR}} \sigma(t)|E(t)|^2 dt \tag{21}$$

The requirements for its value are determined by the dependence of the processing quality indicator on the value of the absorbed energy for an arbitrarily small area within the volume of the processed material, shown in **Figure 7**.

Regardless of the specific goals, microwave EMF processing must ensure heating to a certain temperature; and its insufficient value does not provide the required quality, and its excess leads to the loss of some properties of the processed material. Thus, the requirements for the EMF parameters during Type 1 processing are determined by the relationship:

$$W_{G\min} \leq W\left(\vec{r}\right) \leq W_{G\max} \tag{22}$$

where $W_{G\min}$ and $W_{G\max}$ are respectively the lower and upper limits of the absorbed power values corresponding to the acceptable quality of processing.

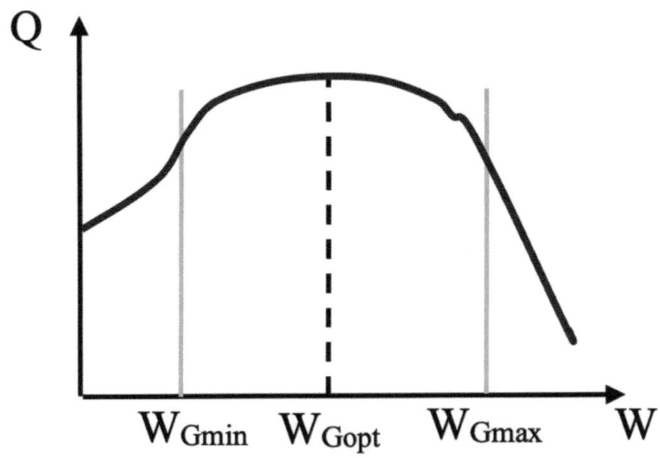


**Figure 7.**
*Typical dependence of processing quality vs. the value of* W.

Type 2 processing is most consistent with the use of specific stimulating effects of UHF EMF exposure on biological objects, as well as, with some reservations, the tasks of inactivating microorganisms during microwave exposure. The nature of the requirements imposed on the EMF parameters to achieve the goals of microwave processing is different in this case.

Type 2 processing has the following characteristic features:

- the effect is achieved when the field intensity exceeds a certain threshold value, and a further increase in the field intensity does not have a noticeable effect on the quality of processing (of course, up to values at which unacceptable heating does not occur)—**Figure 8**;
- the processing time is also characterized by a threshold value, above which the purpose of the processing is ensured, and this value is practically independent of the field intensity—**Figure 9**.

Thus, the requirements for the EMF parameters during Type 2 processing can be formulated as follows: for any point within the processed volume of material $\vec{r} \in V_{обр}$, the following condition for the value of the specific absorbed power of the EMF must be satisfied:

$$P_{G\ \min} \leq P\left(\vec{r}\right) \leq P_{G\ \max} \tag{23}$$

during a given time, interval T

$$T_{G\ \min} \geq T \geq T_{G\ \max} \tag{24}$$

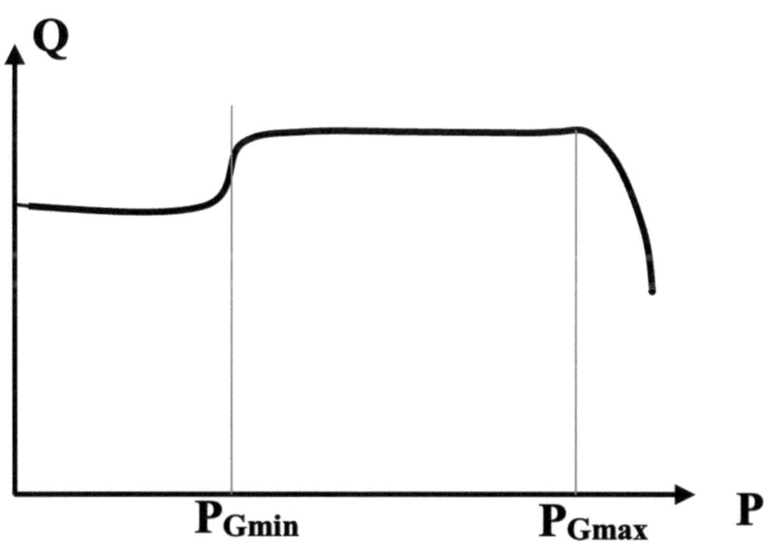


**Figure 8.**
*Type 2 processing, when an increase in the field intensity does not have a noticeable effect on the quality of processing.*

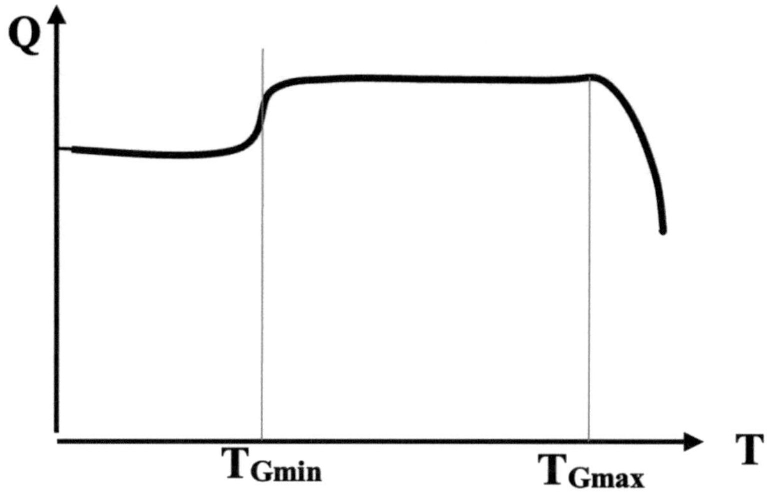


**Figure 9.**
*Type 2 processing, when the quality of processing is practically independent of the field intensity.*

### 4.3 Probabilistic approach

Let us turn to the issue of implementing conditions (21)–(24) in a certain MWTC. Due to the inhomogeneity of the EMF in the working chamber, the specified conditions can be satisfied for some part of the processed material and not for the rest [34–50]. The relationship between the specified parts obviously depends, first of all, on the degree of homogeneity of the EMF. From a practical point of view, the issue of choosing a criterion reflecting the degree of homogeneity of the field and its required numerical value for a particular situation is fundamentally important. The basis for constructing such criteria can be the dependence of the quality achieved during microwave processing, taking into account the heterogeneity of the spatial distribution of the specific absorbed power or energy of the EMF.

For Type 1 processing, as noted above, the value of the specific absorbed energy plays a decisive role. Let the distribution function of this value be known for a certain specific chamber. Let us also assume that the dependence of the quality of microwave processing $Q_I(W)$ is known. An assessment of the quality indicators of processing, taking into account the existing heterogeneity of the field, can be obtained using the concept of the Lebesgue measure. Let $M(W)$ be the value of the specified measure for a set of points in the volume for which the value of the specific absorbed energy corresponds to the interval $[W, W + \Delta W]$. Then the average quality of processing by volume can be defined as:

$$Q_{I\,av} = \int_{0}^{\infty} M(W)\, Q_I(W) dW \tag{25}$$

and the share of material processed with the required quality in the total volume as:

$$\frac{V(Q \geq Q_{ad})}{V} = \int_{W_{G\min}}^{W_{G\max}} M(W) Q_{I}(P) dW \tag{26}$$

Similarly, the quality of Type 2 processing can be assessed. Let $Q_{II}(P,t)$ be a known dependence of the quality index of processing on the EMF intensity and duration of exposure, $M(P,t)$ be a measure of the set in the space of coordinates and time that meets the conditions:

$$P \leq P\left(\vec{r}, t\right) \leq P + \Delta P, T \leq t \leq T + \Delta T$$

Then the average quality of processing by volume of material is determined by the following relationship:

$$Q_{II\ av} = \int_{P_{G\min}}^{P_{G\max}} \int_{T_{G\min}}^{T_{G\max}} M(P,t) dP dt \tag{27}$$

and the relative share of material processed with the required quality:

$$\frac{V(Q \geq Q_{ad})}{V} = \int_{P_{G\min}}^{P_{G\max}} \int_{T_{G\min}}^{T_{G\max}} M(P,t) dP dt \tag{28}$$

In practical assessments, the most natural representation of the measure is in the probabilistic sense. Firstly, for any processed materials, the exact values of the main electro-physical parameters are unknown—the magnitude of electrical permittivity $\varepsilon$ and the loss tangent—$tg\delta$. The inevitable difference in batches of materials (grade, chemical composition, initial humidity) leads to a spread of the specified parameters. As a result, only the average values of the specified parameters and the magnitude of the spread (possible deviations from the average value) can be reliably obtained.

Secondly, during the processing, changes in the specified parameters are possible due to heating, changes in humidity, etc. These changes, in principle, can be calculated with known EMF parameters and processing time. However, for most cases, carrying out these calculations is associated with exceptional difficulties and can hardly be carried out practically with acceptable accuracy. Therefore, only some average values $\varepsilon$ and $tg\delta$, and the limits of possible deviations of these values from the average values are reliable.

Thirdly, the presence of a spread of the initial parameters and their possible change during the processing of the microwave range complicates the already extremely complex task of calculating the spatial distributions of EMF in the volume of processed materials.

The practical impossibility of accurate calculations of the values $W\left(\vec{r}\right)$ and $P\left(\vec{r}\right)$ excludes the direct use of relations (25)–(28) to obtain quantitative estimates of the quality of electromagnetic processing. To overcome these difficulties, it is advisable to use a principle that is widely used in the analysis of physical processes, information about which does not allow their exact description. The essence of this principle is that the quantities of interest are considered random, described by some probabilistic characteristics. In the cases under consideration, it is advisable to consider the absorbed power for a random point in the volume $V$ at a fixed point in time to be a random variable—$\tilde{P}(t)$. This quantity is characterized, in the general case, by the joint distribution density $w(P,t)$. The use of such a representation allows us to answer a number of practically important questions. First, to estimate the probability of implementing the required processing mode for the entire volume $P$ being processed. Indeed, for the case of Type 1 processing

$$P = \int_{W_G}^{\infty} w(W)dW \tag{29}$$

where $w(W)$ is the distribution density of the quantity $W = \int_0^T P(t)dt$.

For the case of the Type 2 processing:

$$p = \int_{P_{G\min}}^{P_{G\max}} \int_{T_{G\min}}^{T_{G\max}} w(p,t)dPdt \tag{30}$$

Second, using the distribution density function of the specific absorbed power, we can estimate the quality of microwave processing with a known dependence on the properties of the product being processed. For Type 1 processing:

$$Q_I = \int_0^{\infty} w(W)Q_I(W)dW \tag{31}$$

Similarly, for Type 2 processing:

$$Q_{II} = \int_0^{\infty} \int_0^{\infty} w(P,t)Q(P,t)dPdt \tag{32}$$

Thus, the use of probabilistic representations allows us to obtain quantitative estimates:

- the probability of implementing processing with specified requirements based on data on field intensity and duration of exposure;
- estimates of the average “quality” of processing taking into account the imperfect fulfillment of the specified processing conditions for all points of the volume being processed;
- formulate requirements for the functions $w(P,t)$ based on the values of the permissible “quality” of processing microwave electromagnetic fields, that is, the permissible values of the quantities $Q_I$ or $Q_{II}$, evaluate their parameters and, on this basis, formulate the necessary requirements for the homogeneity of the generated electromagnetic fields.

### 4.4 Practical examples of type 1 and type 2 thermal processing. Water/oil emulsion separation

Oil hydration (ratio of oil, water, and solid sediments) is one of the most important characteristics for both technological and commercial purposes. All structures involved in the production, processing, and transportation of oil are interested in its flooding, starting from the wellhead and ending with the cargo park.

For the analysis of water-oil emulsion (WOE) composition, a number of devices are proposed, the principle of operation of which is based on preliminary separation of WOE into components (centrifugation) and subsequent measurement of their ratio, as well as spectrophotometric measurements in the IR range. The disadvantage is the low degree of automation of the process and the low accuracy of determining the percentage ratio of the WOE components due to the strong intermolecular bonds of water and oil, and the impossibility of separate determination of the ratio of sediment and water.

Separation of WOE into components occurs under the action of gravitational forces, accelerated by the influence of microwave radiation on the weakening of intermolecular bonds of water and oil at temperatures of 60–90°C. As a result of the separation in the radio and light transparency volume, the structure of the separated WOE is formed as the structure of the video signal received from the charge couple device (CCD).

Next, determine the percentage ratio of the WOE component. According to the change of the amplitude of the video signal, the boundary of the "sediment-water" and "water-oil", as the upper limit of the separated oil, as well as the number of pixels and their ratio, are determined.

The results of EMF distribution in the operation chamber for different *T* and constant P, photography of microwave unit "DON-2450" with monitor for visualization of separated WOE with open and closed microwave generator and few patents for it are shown on **Figure 10A** and **B**.

Type 1 processing is realized in the laboratory stage of WOE separation.

For a moving medium, for example, oil in a pipeline, in an end phase divider channel (**Figure 11A**) for WOE industry exploitation magnetron on 915 MHz is used. Heating WOE in the flow in the pipeline allows using a powerful local source of

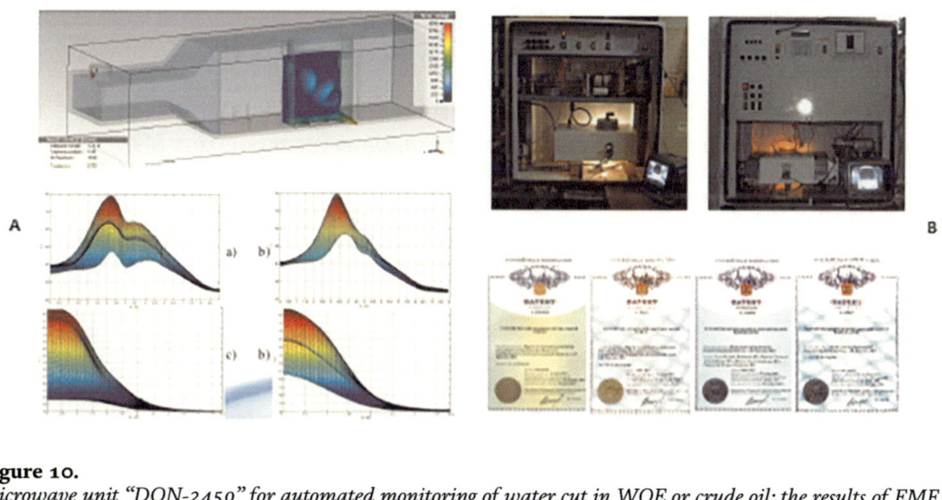

**Figure 10.**
*Microwave unit "DON-2450" for automated monitoring of water cut in WOE or crude oil: the results of EMF distribution in the operation chamber for different* T *and constant* P *(A); photography of microwave unit "DON-2450" with monitor for visualization of separated WOE with open and closed microwave generator and a few patents (B).*

**Figure 11.**
*Industrial microwave crude oil phase divider IMWTC-400: end phase divider channel (A); composite diaphragm for separation SHF energy and WOE (B), block magnetron 915 MHz (C); waveguide output—waveguide of rectangular section 223 × 108 mm² (D).*

microwave energy without the risk of overheating the liquid. At the same time, the introduction of microwave energy into the pipeline should be carried out directly in the zone of liquid flow.

The power of the microwave energy source required to heat the WOE to 1°C, depends on the flow rate of the liquid. Calculation of these parameters for the case of WOE flow rate equal to $\xi = 2\ m^3/h = 0.55\ l/s$ shows P = 2.3 kW. An increase in temperature by 10°C requires power 23 kW.

The scheme of the industrial microwave crude oil phase divider IMWTC-400 is shown in **Figure 11A**. It includes: composite diaphragm for separation SHF energy and WOE (**Figure 11B**), block magnetron 915 MHz (**Figure 11C**); waveguide output—waveguide of rectangular section 223 × 108 $mm^2$ (**Figure 11D**).

Type 2 processing is realized in an industrial microwave crude oil phase divider.

## 5. Conclusion

As noted in Sections 2–4, for the expanded use of MWTC, it is necessary to develop a system of approaches that allow reducing the specific microwave energy consumption, which ultimately leads to a reduction in the costs of implementing MWTC.

The introduction of these levels is dictated by the fact that each of them must solve the general problem of saving microwave energy consumption, but the possibilities for this and, accordingly, the criteria are different.

The choice of the microwave processing mode is currently the prerogative of specialists in the field of materials science, physicists, biologists, etc. The recommended microwave processing mode is a conclusion from the conducted research, moreover, in a specific subject area, and in some sense the best in terms of achieving positive effects (it ensures, e.g., the maximum drying process speed, the best preservation of vitamins during microwave processing compared to traditional methods, the maximum efficiency of microflora inactivation during sterilization of containers, instruments, etc.). At the same time, as a rule, the issue of choosing the frequency range, duration, and other parameters of the EMF from the point of view of the simplicity of hardware implementation, cost of equipment, etc., is not discussed. As a result, for many MWTP, especially those related to the low-intensity group, there are, and sometimes significant, reserves for reducing the cost due to a rational choice of the processing mode.

These results, despite their preliminary nature, are also of some interest from the point of view of reducing the energy costs. Firstly, since the best of the achieved results exceeds the value obtained with separate action, dual-frequency seed treatment may prove more effective from the point of view of the relationship between reduced energy costs and quality. Secondly, the possibility of achieving a better result gives grounds to expect that a result similar to the case of single-frequency treatment may be achieved with lower total reduced energy costs.

In addition to energy savings, improvements can also be achieved in the functional characteristics of complexes and microwave installations for monitoring processed materials.

Compared to the existing methods of analysis, WOE microwave separation provides a higher quality separation of components, and the use of CCD provides a higher accuracy in determining the composition of the components. The error of determining the boundary of the WOE component after its separation with the help of CCD is 1/N, where N is the number of pixels in the line of the video signal. At a standard resolution of 480 pixels, the error is 0.1%. When using the modules, the measurement accuracy increases by 4–5 times compared to similar devices, and an accuracy comparable to precision laboratory devices used in practice in the oil industry (Dina-Stark method) is achieved.

The confidence interval for the mean square deviation of the results of measurements of the immersion of the WOE for the module DON-2450 with a confidence probability of 0.95 is valid within the limits of $0.449 < \Delta < 10.78$, for a laboratory installation implementing the Dina-Stark method, $0.444 < \Delta < 10.66$.

Thus, the confidence interval of the module is included in the confidence interval of the laboratory installation, which implements the Din-Stark method, indicating its high metrological characteristics.

To implement heating in IMWTC-400, it is proposed to use an energy module based on the M-116 magnetron. The power of the microwave module is 50 kW. Frequency is 915 MHz. The generator allows smooth adjustment of output power from 5 to 50 kW. Combined cooling of the generator—water and air. The installation of the generator requires a power supply up to 100 kW, water with a flow rate of 1 $m^3$/h, and compressed air (compressor). The installation allows to increase the temperature of the flow of oil with flow rate $\xi = 4$ $m^3$/h by 10–15°C.

## Notes

This chapter is devoted to the jubilee of Kazan microwave technologies science school founder, Gennady A. Morozov.

## Author details

Gennady Morozov, Yuri Sedelnikov, Vladimir Anfinogentov, Natalya Stakhova and Oleg Morozov*
Kazan National Research Technical University n.a. A.N. Tupolev-KAI, Kazan, Russia

*Address all correspondence to: ogmorozov@kai.ru

IntechOpen

© 2025 The Author(s). Licensee IntechOpen. This chapter is distributed under the terms of the Creative Commons Attribution License (http://creativecommons.org/licenses/by/4.0), which permits unrestricted use, distribution, and reproduction in any medium, provided the original work is properly cited.

## References

[1] Osepchuk JM. A history of microwave heating applications. IEEE Transcription On Microwave Theory and Techniques. 1984;**MTT-32**(9):1200-1223

[2] Morozov GA, Morozov OG, Stahova NE, Stepanov VV, Sedelnikov YE. Low-Intensity Microwave Technology: Methods and Apparatus. Moscow: Radioelectronics and Communications Systems; 2003. 128 p. (in Russian)

[3] Thuery J's. In: Grant EM, editor. Microwave Industrial. Scientific and Medical Applications. Boston, London: Larstin Arteda House; 1992

[4] Vakhin AV, Khelkhal MA, Mukhamatdinov II, et al. Changes in heavy oil saturates and aromatics in the presence of microwave radiation and iron-based nanoparticles. Catalysts. 2022;**12**(5):514

[5] Vakhin AV, Khelkhal MA, Tajik A, et al. The role of nanodispersed catalysts in microwave application during the development of unconventional hydrocarbon reserves: A review of potential applications. PRO. 2021;**9**(3):1-20

[6] Danilaev MP, Bogoslov EA, Morozov OG, et al. Obtaining carbon dendrites from the products of conversion of polymer materials. Journal of Engineering Physics and Thermophysics. 2015;**88**(3):774-780

[7] Kuklin V, Karandashov S, Bobina E, et al. Analysis of aluminum oxides submicron particle agglomeration in polymethyl methacrylate composites. International Journal of Molecular Sciences. 2023;**24**(3):2515

[8] Vedenkin D, Morozov O, Sedelnikov Y, et al. Increasing the accuracy characteristics of focused electromagnetic devices for non-destructive testing and technical diagnostics by implementing sum–difference signal processing. Electronics. 2023;**12**(2):436

[9] Jolly JA, Tate RL. Douglas fir tree seeds germination enchacement using Microwave energy. Journal of Mi-Crowave Power. 1971;**6**(2):125-130

[10] Nelson SO et al. Long term effects of RF dielectric heating on germination of alfa alfa seeds. Transactions of the ASAE. 1984;**27**(1):0255-0258

[11] Nelson SO, Stretson LE. Germination responses of selected plants species to RF electrical seeds treatment. Transactions of the ASAE. 1985;**28**(6):2051-2058

[12] Tran VN. Optimising of the microwave treatment of acacia seeds. Journal of Microwave Power. 1981;**16**(3): 277-281

[13] Morozov GA, Morozov OG, Samigullin RR, Nasibullin AR. Application of microwave technologies for increase of efficiency of polymeric materials recycling. In: Proceedings of IEEE 8th International Conference on Antenna Theory and Techniques. IEEE; 2011. pp. 321-323

[14] Anfinogentov VI, Morozov GA, Stakhova NE, Smirnov SV, Ganieva SR. Optimization of microwave heating of dielectrics taking into account errors in amplitudes of the excitation electromagnetic field emitters. In: Proceedings of IEEE 11th International Conference on Antenna Theory and Techniques. IEEE; 2017. pp. 433-436

[15] Morozov G, Anfinogentov V, Morozov O, Smirnov S, Ganieva S.

Modeling and optimization of microwave heating in cylindrical volumes. In: Proceedings of IEEE 11th International Conference on Antenna Theory and Techniques. IEEE; 2017. pp. 68-73

[16] Smirnov SV, Nasybullin AR, Morozov GA, Anfinogentov VI, Lipatnikov KA. Mathematical model for technological process of organic livestock waste microwave treatment in conveyor installation. In: Proceedings of IEEE Wave Electronics and its Application in Information and Telecommunication Systems. IEEE; 2021. p. 9470652

[17] Morozov GA, Shangaraeva YN. Microwave heating. The design, modeling and monitoring of thermal processes and complexes. In: Proceedings of IEEE 9th International Conference on Antenna Theory and Techniques. IEEE; 2013. pp. 526-528

[18] Chetverikov AP, Yavchunovsky VV, Morozov GA, Yakunin AN. The modeling of electrodynamics and thermodynamics processes while electromagnetic field and dielectric interaction. In: Proceedings of IEEE 5th International Conference on Antenna Theory and Techniques. IEEE; 2005. pp. 491-493

[19] Morozov GA, Anfinogentov VI, Galimov MR, Morozov OG. Mathematical modeling of the UHF heating of multilayered liquid in static cylindrical volume. In: Proceedings of IEEE 5th International Conference on Antenna Theory and Techniques. IEEE; 2005. pp. 494-496

[20] Morozov GA, Morozov OG. Mathematical modeling of the UHF heating of water oil emulsions. In: Proceedings of IEEE 5th International Conference on Antenna Theory and Techniques. IEEE; 2005. pp. 99-104

[21] Anfinogentov VI, Morozov GA. The review of methods of physical and mathematical modelling of the microwave heating. In: Proceedings of IEEE 13th International Crimean Conference "Microwave and Telecommunication Technology". IEEE; 2003. pp. 710-711

[22] Morozov GA, Morozov OG. Development of microwave processes and complexes today. In: IEEE 2015 International Conference on Antenna Theory and Techniques (ICATT). 2015. pp. 1-6

[23] Gusev VF, Abrajeev GP, Galimov RC, Morozov GA, Vorobiev NG. Microwave technologies application for oil shlamm fusion. In: Proceedings of IEEE 9th International Crimean Microwave Conference: Microwave and Telecommunication Technology. IEEE; 1999. pp. 414-415

[24] Gusev VF, Yakimov IM, Kalachev IF, Galimov RC, Morozov GA. Processes modeling in microwave complexes of oil and gas industry. In: Proceedings of IEEE 9th International Crimean Microwave Conference: Microwave and Telecommunication Technology. IEEE; 1999. pp. 416-417

[25] Anfinogentov VI, Garaev TK, Morozov GA. Increase of uniformity of temperature fields at the microwave heating in open chambers. In: Proceedings of IEEE 13th International Crimean Conference "Microwave and Telecommunication Technology". IEEE; 2003. pp. 714-715

[26] Morozov GA, Morozov OG, Galimov MR. Development of microwave technologies for the oil and gas extraction complex. In: Proceedings of

IEEE 4th International Conference on Antenna Theory and Techniques. IEEE; 2003. pp. 101-104

[27] Morozov OG, Morozov GA. Microwave techniques for gas-and-oil producing industry. In: Proceedings of IEEE 12th International Conference "Microwave and Telecommunication Technology". IEEE; 2002. pp. 28-31

[28] Morozov GA, Vorobiev NG, Salikhov AM, Korpatchev YA. Experimental simulation of power performances of Tyre pyrolysis. In: Proceedings of IEEE 12th International Conference "Microwave and Telecommunication Technology". IEEE; 2002. pp. 563-564

[29] Anfinogentov VI, Garayev TK, Morozov GA. Optimization of dielectric microwave heating by moving radiator. In: Proceedings of IEEE 12th International Conference "Microwave and Telecommunication Technology". IEEE; 2002. pp. 605-606

[30] Morozov GA. Microwave technological complexes design in view of power consumption factor. In: Proceedings of IEEE 10th International Crimean Microwave Conference "Microwave and Telecommunication Technology". IEEE; 2000. pp. 61-62

[31] Kappe CO. How to measure reaction temperature in microwave-heated transformations. Chemical Society Reviews. 2013;**42**(12):4977-4990

[32] Herrero MA, Kremsner JM, Kappe CO. Non-thermal microwave effects revisited: On the importance of internal temperature monitoring and agitation in microwave chemistry. The Journal of Organic Chemistry. 2008;**73**(1):36-47

[33] Morozov GA, Morozov OG. On-line NDE monitoring for microwave reticulation of thermosetting resins. Proceedings of SPIE. 1998;**3396**:264-270

[34] Nasybullin AR, Morozov GA, Anfinogentov VI, Lipatnikov KA, Smirnov SV. Multi-sensor temperature and humidity measuring system for technological process of organic livestock waste microwave treatment monitoring. In: Proceedings of IEEE Wave Electronics and its Application in Information and Telecommunication Systems. IEEE; 2021. p. 9470606

[35] Vorobiev NG, Galimov MR, Dorogov NV, et al. Microwave modules for watering monitoring of wateroil emulsions. In: Proceedings of IEEE 13th International Crimean Conference "Microwave and Telecommunication Technology". IEEE; 2003. pp. 712-713

[36] Morozov OG, Morozov GA. Television methods for measurement systems with microwave visualization of monitoring objects. In: Proceedings of IEEE 11th International Conference "Microwave and Telecommunication Technology". IEEE; 2001. pp. 623-625

[37] Dorogov NV, Galimov MR, Galimov RG, et al. Experimental sets for process monitoring of water-oil emulsion decomposition during microwave heating. In: Proceedings of IEEE 10th International Crimean Microwave Conference "Microwave and Telecommunication Technology". IEEE; 2000. pp. 63-64

[38] Li J, Xiong Q, Wang K, Shi X, Liang S, Gao M. The application of MPC in microwave heating process based on model constructed by Lambert's law combined with temperature. Mathematical Problems in Engineering. 2015;**2015**:610490

[39] Pitchai K, Chen J, Birla S, Gonzalez R, Jones D, Subbiah J. A microwave heat

transfer model for a rotating multi-component meal in a domestic oven: Development and validation. Journal of Food Engineering. 2014;**128**:60-71

[40] Geedipalli SSR, Rakesh V, Datta AK. Modeling the heating uniformity contributed by a rotating turntable in microwave ovens. Journal of Food Engineering. 2007;**82**(3):359-368

[41] Ayappa KG. Resonant microwave power absorption in slabs. Journal of Microwave Power and Electromagnetic Energy. 1999;**34**(1):33-41

[42] Bhattacharya M, Basak T. A novel closed-form analysis on asymptotes and resonances of microwave power. Chemical Engineering Science. 2006;**61**(19):6273-6301

[43] Basak T. Analysis of resonances during microwave thawing of slabs. International Journal of Heat and Mass Transfer. 2003;**46**(22):4279-4301

[44] Basak T, Rao BS. Theoretical analysis on pulsed micro-wave heating of pork meat supported on ceramic plate. Meat Science. 2010;**86**(3):780-793

[45] Huang K, Lu B. The precise condition of thermal runaway in microwave heating on chemical reaction. Science in China, Series E: Technological Sciences. 2009;**52**(2):491-496

[46] Fan DM, Li CX, Ma WR, Zhao JX, Zhang H, Chen W. A study of the power absorption and temperature distribution during microwave reheating of instant rice. International Journal of Food Science and Technology. 2011;**47**(3): 640-647. DOI: 10.1111/j.1365-2621.2011.02888.x

[47] Rattanadecho P. Theoretical and experimental investigation of microwave thawing of frozen layer using a microwave oven (effects of layered configurations and layer thickness). International Journal of Heat and Mass Transfer. 2004;**47**(5):937-945

[48] Campanone LA, Zaritzky NE. Mathematical modeling and simulation of microwave thawing of large solid foods under different operating conditions. Food and Bioprocess Technology. 2010;**3**(6):813-825

[49] Hemis M, Singh CB, Jayas DS. Microwave-assisted thin layer drying of wheat. Drying Technology. 2011;**29**(10): 1240-1247

[50] Baer C, Orend K, Hattenhorst B, Musch T. Field representation microwave thermography utilizing lossy microwave design materials. Sensors. 2021;**21**:4830

Chapter 8

# Static Magnetic Fields and Cancer Therapy

*Joseph R. Salvatore and Andrea Bailey*


## Abstract

Magnetic fields, static or other types, have a long history of use as a healing therapy for Human medical conditions. Medical and scientific research has shown that, within certain parameters, magnetic fields have diagnostic and therapeutic applications in Human medicine. The majority of these therapeutic applications are for medical conditions other than cancer, and static magnetic fields are used in diagnostic medical applications (e.g., Magnetic Resonance Imaging). There are studies, both cell culture and animal models, that have applied static and other magnetic fields to determine if the fields are effective alone as an antineoplastic or if, in combination, they increase the effectiveness of antineoplastic chemotherapy. Human studies have been conducted and designed to at least determine if magnetic fields are safe to apply to patients with cancer diagnoses, including those receiving antineoplastic chemotherapy, and some show favorable results. We review some of the laboratory and clinical studies done to determine if static magnetic fields, with or without antineoplastic chemotherapy, can be effective as cancer therapy, and if static magnetic fields are safe to apply in clinical medicine.


**Keywords:** magnetic fields, static magnet, magnet, static magnetic fields, cancer, tumor, antineoplastic, cancer therapy

## 1. Introduction

The concept that magnets have a healing power has a long history in public thought. In popular and medical literature, there are both unsubstantiated claims of the healing power of magnets and many rigorous studies, both laboratory and clinical, that show the actual and potential for magnets and magnetic fields to be a successful diagnostic and therapeutic medical intervention. There are available reviews of the history of magnetism in medicine, and an extensive review is beyond the scope of this chapter [1, 2].

In the London Medical Journal in 1782, Thomas Henry reported two cases of dental pain and one of tinnitus being relieved by the application of magnets [3]. Interestingly, it was Henry himself who had one of the cases of dental pain and the tinnitus, and his self-administered magnets provided successful therapy for both illnesses.

In the Lancet in 1880, Dr. Allison published a case report describing the "therapeutic effects of lightning upon cancer" [4]. The case report describes a farm worker with a malignancy of the lip/chin who became incapacitated after being struck by lightning and when examined by Dr. Allison, was "emitting a strong odor of ozone." The neoplasm was observed to regress completely following this injury. Although the electric current induced by the lightning is the presumed physical force that caused the antineoplastic regression, intrinsic induced ionic flow from the lightning may have generated a static magnetic field that also caused the regression. Changes in cell membrane potential have been related to control or lack of control of cellular mitosis, resulting in neoplastic change [5–7]. Other theories of the relationship between electronics, biology, and cancer have been published [8].

The use of static magnetic fields (smf) in Human medical therapy has focused more on medical diagnoses other than malignancy. In particular, smf therapy has been used for Orthopedic and Rheumatologic disorders, wound healing, neuropathy, and the therapy of pain [9–15].

## 2. Static magnetic fields

In understanding the effects of an smf, the concept of Magnetic Flux Density (MFD) provides a generally accepted measurement of the magnetic force potentially related to the effects of the smf [16]. For smf to exert a force on a charge, that charge requires movement, such as ionic charges in blood flow [17]. The terms MFD and Magnetic Field (MF) are both used to represent the force from an MF [18].

The units of MF measurement, Tesla and Gauss, represent a quantity of MF strength and are related as 1Tesla (T) = $10^4$ Gauss (G). In this chapter, we will use Tesla to describe MF's strength.

When considering the penetration of magnetic fields into and through biological tissue, the permeability of tissue is similar to that of free space; there is a relative equivalence of the field strength at the origin of the MF and in the target tissue/organ [19]. Others have pointed out that body tissue does not attenuate smf [20].

Smf is a natural Earth phenomenon, man-made and intrinsic to Human physiology. The Earth produces a natural smf due to the presence of Iron and other metals in core areas. Although more complicated than this simplistic model, Iron moving in parts of the Earth's core is essentially a moving electrical current that produces an smf, and measurements of the strength of this smf show variation depending on where on Earth the measurement is taken [21, 22]. Measurement of Earth's MF shows a range of MF strength from 25 to 65 uT (varying by location on the Earth) and the MF lines move from the Southern Pole to the Northern Pole [23]. It is generally understood that all life forms on Earth have likely been exposed to this natural MF since life began. There is data suggesting that manipulation of this Geomagnetic field can produce change in the results of Biological and Medical investigations [24]. The existence and measurement of this Geomagnetic field should be considered when conducting smf research. Also, natural MF is generated by ionic current flow in the Human body which is very low level and requires special equipment to detect. The MF produced by the Heart are orders of magnitude smaller than even the Earth's MF (Heart: 50–100 pico Tesla (pT; pico = $10^{-12}$) vs. Earth: 30,000,000 pT), and this represents a challenge to the detection of the Heart's MF by Magnetocardiogram [25]. There are similar challenges to the detection of the MF of the Brain by Magnetoencephalogram.

| Application | Frequency |
|---|---|
| **Non-ionizing** | |
| Static magnetic field | 0 Hertz |
| Electric power | 60 Hertz (U.S.) |
| Radio frequency | $10^4$ Hertz |
| Microwaves | $10^{10}$ Hertz |
| **Ionizing** | |
| Ultraviolet | $10^{16}$ Hertz |
| X-rays | $10^{18}$ Hertz |
| Gamma rays | $10^{20}$ Hertz |

**Table 1.**
*Relationship between application and approximate frequency.*

Of the important physical characteristics of smf is the concept that these fields and their MFD, do not vary over time (hence static or direct). This is as opposed to MF produced by an alternating current where the direction of the MF does vary ("alternates") with time.

The use of a static magnet in Human medical applications accepts a few basic concepts. The MF strength with these permanent magnets is fixed, and changes in the applied field strength would be accomplished changing the strength of the MF, by varying the distance to the target (e.g., Knee or Liver), by placing something between the magnet and the target, or by changing the number and placement of the magnets. Smf has been applied to cancer cell culture systems and animal models to which have been added a number of antineoplastic chemotherapeutic agents (anc). These studies have been done to determine if there is a modifying-modulating effect, or none, by the smf on the interaction between the cell culture or animal models and anc, in particular, whether there is a change in the efficacy of the anc. A limited number of clinical studies have been conducted.

It is clear that Radiation (e.g., UltraViolet Radiation and X-Rays, in general > a Frequency of $10^{15}$ Hertz) is carcinogenic (**Table 1**). However, although smf of 0 Hertz may be able to affect different subcellular functions, there is no data to suggest that these fields are directly carcinogenic.

## 3. Cancer therapy: SMF and laboratory studies

Considerable non-clinical research has been done using cell culture and animal models, applying an smf with or without anc. There are a number of mechanisms that may be responsible for the observed effects of smf as a solo agent or in combination with anc to alter the growth of cancer cells in culture or tumors in animal models. **Table 2** lists some of the proposed mechanisms. **Table 3** reviews the types of cell lines used in the cell culture laboratory research.

Smf laboratory studies have been conducted with very low and very high MF strengths, some many times higher than studies that apply small magnets directly to the skin.

| |
|---|
| **Apoptosis** |
| Cell membrane changes |
| Decreased neoplastic cell metastatic potential |
| Decreased neoplastic cell migration |
| Effects on ions in blood flow |
| Effects on telomerase function |
| **Genotoxicity** |
| Increased reactive oxygen species with oxidative stress |
| Increased microvascular permeability |
| Inhibition of tumor angiogenesis |
| Interference with the mitotic spindle |
| Reduced drug resistance |

**Table 2.**
*Possible mechanisms of action of static magnetic fields in biological/human systems.*

| Cell line name | Cell type |
|---|---|
| MCF-7 | Female breast cancer |
| K562 | Hematopoietic cell from chronic myelocytic leukemia in blast crisis. Likely an erythroleukemic cell [26]. |
| A2780 | Ovarian cancer cell |
| A2780CP | Ovarian cancer cell resistant to cisplatinum chemotherapy |
| SH-SY5Y | Neuroblastoma cell line |
| HeLa | Cervical cancer cells |
| SK-OV-3 | Ovarian adenocarcinoma |
| HO-8910 | Originally considered an ovarian cancer cell line, but may actually be a derivative of HeLa cell line [27]. |
| HCT116 | Colon cancer |
| CNE-2Z | Originally considered to be a nasopharyngeal cell line. May be a hybrid of HeLa and an unknown cell line [28]. |
| CHO with EGFR overexpression | Chinese hamster ovary cells with epidermal growth factor receptor. EGFR overexpression is associated with cancer progression [29]. |
| 16/C | Murine mammary adenocarcinoma |
| 4 T1 | Breast cancer cells derived from a mouse model |

**Table 3.**
*Neoplastic cell lines mentioned in this chapter.*

A cancer cell culture of 4 T1 cells was exposed to an smf and Telomerase function and cell migration, both likely part of the ability of malignant cells to proliferate and metastasize, were measured [30]. This work demonstrated that both can be inhibited by an smf. Chinese Hamster Ovary (CHO) cells with Epidermal Growth Factor Receptor (EGFR) overexpression were exposed to a 9 T magnet. The result showed inhibition of proliferation of the overexpressed EGFR cells, but CHO cells without the overexpression showed much less inhibition [31]. Using a superconductivity magnet of 9.4 T, another research group placed Human Ovarian cancer cells (SK-OV-3 and HO-8910) in an area of the magnet that was ~0.5 T. In addition to the superconductivity magnet, a magnetic plate with 12 neodymium magnets was used which also had a MF strength of 0.5 T [32]. The authors indicated that the results of their study, a decrease in ovarian cancer cell migration and metastasis, was likely related to an increase in reactive oxygen species (ROS) induced by the MF application.

A 1 T smf was found to interfere with the function of microtubules and mitotic spindles. This group exposed a series of neoplastic cell lines (MCF-7, HeLa, HCT116, CNE-2Z) and three anc agents, fluorouracil, paclitaxel, and cisplatinum, along with the smf [33]. The combination of smf and fluorouracil or paclitaxel showed an improvement in antineoplastic activity over the agents alone, but this was not seen using cisplatinum. It should be noted that a mechanism of action of paclitaxel is to interfere with microtubules. Another study using multiple anc agents in combination with an smf of 9mT found changes in the cell membrane as the likely mechanism for enhanced antineoplastic cell action [34]. This study did not vary the MF strength (9mT variation <1%) and used only one cell line – K562. Only the anc agents were multiple and included cyclophosphamide, doxorubicin, paclitaxel, and cisplatinum.

The cell lines A2780 and A2780 CP were exposed to a combination of cisplatinum anc with smfs of 10, 15, and 25 mT [35]. The results of this study suggested that the addition of the smf decreased neoplastic cell resistance to cisplatinum. This paper also notes that cisplatinum is a diamagnetic substance.

A follow-up study by this group used an smf of ~15 mT applied to the same cell culture system with and without the addition of cisplatinum [36]. The results of this study showed an increase in the effectiveness of the cisplatinum when combined with the smf for both cell lines and the researchers indicated decreased resistance to anc and an increase in cell cycle inhibition and apoptosis as the likely mechanism for the observed result.

K562 neoplastic cells were exposed to a combination of cisplatinum, anc, and an smf of 8.8mT, and the researchers reported enhanced destruction of the neoplastic cells by the combination therapy [37].

An increase in oxidative stress was noted in HeLa cells exposed to a combination of cisplatinum and smf intensities of 7, 10, and 15 mT [38]. The authors report that the smf of 10 mT was most effective in sensitizing the HeLa cells to the cisplatinum anc. Another study that used a 10mT smf combined the smf with the anc doxorubicin against an MCF-7 cell culture system [39]. These researchers observed a decrease in the proliferation of these neoplastic cells exposed to the combined therapy and noted that there was an increase in the production and persistence of ROS.

Animal studies have also been conducted using smf and anc agents. One animal study combining an smf of 587 mT and paclitaxel showed increased the effectiveness of the anc [40]. In addition to paclitaxel exerting its antineoplastic activity by binding to microtubules resulting in cell cycle arrest, the researchers indicated that there was inhibition of tumor angiogenesis and increased tumor microvessel permeability.

In another animal study, mice were implanted with a 16/C neoplastic cell line [20]. Following the growth of the implanted neoplasm, one of the four animal groups was exposed to a combination of the anc doxorubicin and an smf of 110 mT. The researchers reported that the group exposed to this combination showed a statistically significant tumor regression.

Not all combinations of smf and cisplatinum have shown a synergism toward improvement. Using SH-SY5Y cell culture, researchers applied the anc cisplatinum

| Agent | Mechanism of action |
|---|---|
| Cisplatinum | Crosslink of purine DNA bases resulting in DNA damage |
| Cyclophosphamide | Alkylating agent – nitrogen mustard type agent. Alkylation of DNA |
| Doxorubicin | Intercalation of DNA. Also inhibits DNA repair by interaction with topoisomerase II enzyme |
| Fluorouracil | Antimetabolite – interferes with DNA replication |
| Paclitaxel | Binding to microtubules with stabilization of microtubules |

**Table 4.**
*Mechanisms of action of antineoplastic pharmaceuticals mentioned in this chapter.*

with an smf of 31.7–232mT and found that the cytotoxicity of the cisplatinum was prevented by this therapeutic intervention [41]. Although it is not clear how to interpret these results, we do note that the choice of neoplastic cells for this work, a Neuroblastoma, is somewhat different than other studies.

An in vitro study that used an smf of 1.5 T but no cell culture or animal model explored the effect of the smf on the viscosity of blood. This group found that the smf increased the viscosity of the blood [42]. Although it is not clear if this finding has direct applicability to the therapy of malignancy, it is known that patients with malignancy have a higher incidence of thrombo-embolic disease, and based on these results, exposure to MF may increase this risk [43].

The studies referred to in this chapter using a combination of smf and anc, use a number of originally developed anc agents, many of which remain in standard use today as part of antineoplastic therapy protocols. Results of studies using these agents with smf can be applied to many current cancer therapy protocols. **Table 4** reviews the mechanism of action of those antineoplastic agents.

## 4. Cancer therapy: SMF and clinical cancer studies

Compared to the number of available laboratory studies using smf with or without anc in cancer cell culture of animal cancer models, there are fewer clinical studies that have applied smf with or without anc in patients diagnosed with cancer. Of four clinical studies reviewed here, only one used a pure smf with anc, and the others used a modified smf. Two clinical studies used a modified smf without anc. One of these studies applied a modified smf to 13 patients with advanced stage Non-Small Cell Lung cancer without anc [44]. The researchers observed a palliation of symptoms in many of the patients, including importantly in patients with Lung Cancer, a decrease in Pleural Effusions, an improvement in respiratory status, and relief of pain.

Another group of patients with heavily pre-treated advanced cancers were exposed to an smf which was modified by an extremely low-frequency amplitude modulated signal without anc [45]. The purpose of this study was to determine if this MF could be safely applied to this group of patients. The researchers determined that, using multiple follow-up parameters, the MF in the manner given was safe to use as a therapeutic and clinical research modality.

Patients with advanced Lung cancer were the subjects of the investigation that used a modified smf during the administration of anc [46]. The MF was a 0.4 T smf that was rotating at a frequency of 7.5 Hertz. The study included a control group that was receiving anc but exposed to a sham MF. The researchers measured quality of life (QOL) as well as parameters such as progression-free survival (PFS) and found that there was an improvement in QOL in the MF therapy group compared to the control group, but without evidence of a difference in PFS.

Our clinical study applied a permanent neodymium magnet, non-modified, to patients with advanced cancer on anc. We applied to the magnet during the time of administration of the anc. Then for each of three groups of patients, the magnet was left in place following completion of anc for 0, 15, and 30 minutes. Thirty-nine patients were recruited into the study, and each research subject was paired with a matched control. We are currently reviewing the data from both the initial completion of the study and 20 year follow-up. Our preliminary findings are similar to the other studies in this chapter with the smf-exposed group showing a decrease in side effects related to the anc and an improved QOL. We have not yet determined if the application of the smf affected PFS or Overall Survival (OS). We did publish early data that showed that the application of the smf with anc could be safely administered to patients with advanced cancer [47].

Based on the results of these studies, it can be concluded that, within certain parameters, the application of an smf with or without anc in clinical medical research can be considered safe for further clinical investigation.

## 5. Discussion

A standard question that must be resolved when conducting clinical medical research is to determine the level of risk the research protocol represents to participants, and whether that risk would interfere with the completion of the investigation. Additionally, clinical researchers must be able to identify individuals who are appropriate for a clinical research protocol through specified inclusion and exclusion criteria. There must also be a level of willingness on the part of potential candidates to participate in clinical research, and determining this is often one of the most difficult aspects of clinical research. When our group conducted a survey study to determine the level of magnet use in our patient population, an additional question was if the survey participant would consider being a candidate for clinical research that would use magnets as a major focus of that research. Almost all of the just over 200 survey respondents indicated that they would welcome the opportunity to participate [48].

Although laboratory data provides evidence that smf with or without anc can affect the growth of malignant cells in cell culture or in animal cancer models, clinical evidence for the efficacy of these therapeutic modalities is still being developed but has some promising aspects. Based on current data, it is not possible to conclude that using smf as an adjuvant to anc is an effective antineoplastic therapy. However, if current clinical data shows that smf can reduce the side effects of anc and/or improve a patient's QOL, further clinical research is justified to both confirm those findings and also determine if there is an improvement in the efficacy of anc (i.e., improved PFS or OS). With the development of newer antineoplastic therapies, particularly Biologic and Immunotherapy, clinical research should extend to those modalities. All of this clinical research may be particularly important in patients who have locally advanced or metastatic cancer, have therapy protocols using multiple therapeutic agents (including Radiation Therapy), and patients who have had a series of therapeutic interventions that have become progressively more difficult to tolerate. If the data confirms, at a minimum, that the application of a small permanent magnet(s) can effectively reduce the side effects of anc or palliate symptoms, patients would have access to a therapy that is easily self-administered in partnership with their health care providers. However, any magnetic therapy that would prove useful would be welcomed. Magnets that have been used in most medical therapies are small and can be applied to single or multiple areas of the body. Since there is no power source attached to the magnet, its position can be easily reoriented and the number of magnets can be increased or decreased. This also provides an element of simple portability. Static magnets applied in Human research or therapy need to be close to or in contact with the skin or the target organ to ensure that interaction will take place with the smf, keeping in mind the change in the strength of the smf with increasing distance to the target organ. As previously noted, there is minimal, if any attenuation of the MF by body tissue. Another issue for permanent magnets is that the MF strength is fixed. Some of the ways to deal with this issue would be by changing the magnet's location (i.e. – distance to target), attenuating the magnetic field, or using magnets in an additive manner. Positioning the magnet and knowing the MF strength at specified distances from the magnet would appear to be a practical approach, which was part of our work. There are pros and cons to applying small permanent magnets in medical research/therapy (**Table 5**).

Also recommended for future studies are those focused on oral antineoplastic therapies and modern immunotherapies. However, it should be noted that, because of the differences in mechanism of action between traditional chemotherapy agents and the newer immunotherapies, results of studies with traditional chemotherapies would not necessarily predict the outcome of studies on immunotherapies. Therefore, in vivo studies are more likely to provide the data required to understand the interaction between smf and these newer therapies.

Further investigation of the use of smf in the therapy of cancer are needed at multiple levels. There are a number of proposed mechanisms to account for the possible effects of smf; these would be targets of future investigation. Confirmation that the application of static magnets, or other types of MFs during cancer therapy, may reduce the side effects experienced by patients during anc, and/or improve QOL,

| PROS | CONS |
|---|---|
| Small size | Maintaining location may be difficult |
| Simplicity | Magnetic field strength fixed |
| Does not require a power source | Overall magnetic field strength low |
| Easily relocated on body | Best placement of magnet on the body not yet known |
| **Vary distance to target** | |
| Easily increase or decrease the number of magnets | |
| Easy to use in the outpatient setting | |
| Low risk | |
| Low cost | |
| Outpatient – in home – application | |
| Portable | |

**Table 5.**
*PROS/CONS of direct application of static magnets in cancer therapy.*

is one area of research. Another is to determine if magnets as a direct or adjuvant therapeutic intervention have a beneficial or detrimental effect on the outcome of cancer therapy.

## Author details

Joseph R. Salvatore*† and Andrea Bailey
Division of Hematology/Oncology, VA Medical Center, Phoenix, AZ, USA

*Address all correspondence to: jrs9928@yahoo.com

†(RET.) Affiliation for identification only

**IntechOpen**

© 2025 The Author(s). Licensee IntechOpen. This chapter is distributed under the terms of the Creative Commons Attribution License (http://creativecommons.org/licenses/by/4.0), which permits unrestricted use, distribution, and reproduction in any medium, provided the original work is properly cited.

## References

[1] Basford JR. A historical perspective of the popular use of electric and magnetic therapy. Archives of Physical Medicine and Rehabilitation. 2001;**82**:1261-1269

[2] Mourino MR. From Thales to Lauterbur, or from the lodestone to MR imaging: Magnetism and medicine. Radiology. 1991;**180**:593-612

[3] Henry T. Instances of the medicinal effects of magnetism. London Medical Journal. 1782;**3**(3):303-305

[4] Allison A. Therapeutic effects of lightning upon cancer. Lancet. 1880;**1**:77

[5] Cone CD Jr. Maintenance of mitotic homeostasis in somatic cell populations. Journal of Theoretical Biology. 1971;**30**(1):183-194. DOI: 10.1016/0022-5193(71)90043-9

[6] Cone CD Jr. Unified theory on the basic mechanism of normal mitotic control and oncogenesis. Journal of Theoretical Biology. 1971;**30**(1):151-181. DOI: 10.1016/0022-5193(71)90042-7

[7] Cone CD Jr. Variations of the transmembrane potential level as a basic mechanism of mitosis control. Oncology. 1970;**24**(6):438-470. DOI: 10.1159/000224545

[8] Szent-Gyhorgyi A. Electronic Biology and Cancer. A New Theory of Cancer. NY: Marcel Dekker; 1976

[9] Shi P, Gao H, Cheng Z, et al. Static magnetic field-modulated mesenchymal stem cell-derived mitochondria-containing microvesicles for enhanced intervertebral disc degeneration therapy. NanoBiotechnology. 2024;**22**(1):457. DOI: 10.1186/s12951-024-02728-6

[10] Wang S, Liu Y, Lou C, et al. Moderate static magnetic field promotes fracture healing and regulates iron metabolism in mice. Biomedical Engineering Online. 2023;**22**(1):107. DOI: 10.1186/s12938-023-01170-3

[11] Xiao Y, Shen Q, Li W, et al. 280 mT static magnetic field promotes the growth of postpartum condylar cartilage. Connective Tissue Research. 2023;**64**(3):248-261. DOI: 10.1080/03008207.2022.2148527. Epub 2022 Dec 5

[12] Segal NA, Toda Y, Huston J, et al. Two configurations of static magnetic fields for treating rheumatoid arthritis of the knee: A double-blind study. Archives of Physical Medicine and Rehabilitation. 2001;**82**(10):1453-1460. DOI: 10.1053/apmr.2001.24309

[13] Weintraub MI, Wolfe GI, Barohn RA, et al. Static magnetic field therapy for symptomatic diabetic neuropathy: A randomized, double-blind, placebo-controlled trial. Archives of Physical Medicine and Rehabilitation. 2003;**84**(5):736-746. DOI: 10.1016/s0003-9993(03)00106-0

[14] Feng C, Yu B, Song C, et al. Static magnetic fields reduce oxidative stress to improve wound healing and alleviate diabetic complications. Cells. 2022;**11**(3):443. DOI: 10.3390/cells11030443

[15] Fan Y, Ji X, Zhang L, et al. The analgesic effects of static magnetic fields. Bioelectromagnetics. 2021;**42**(2):115-127. DOI: 10.1002/bem.22323. Epub 2021 Jan 28

[16] ICNIRP. Guidelines on limits of exposure to static magnetic fields. Health Physics. 2009;**96**(4):509-514

[17] Edmonds DT. Electricity and Magnetism in Biological Systems. Chapter 10: The Motion of a Charged Particle in Electric and Magnetic Fields and Relativity. Great Britain: Oxford University Press; 2001. pp. 158-159

[18] Nagel JR, Furse CM, Christensen DA, et al. Chapter 1: Electric and magnetic fields. Basic concepts. In: Nagel JR, Furse CM, Christensen DA, et al., editors. Basic Introduction to Bioelectromagnetics. 3rd ed. Boca Raton: CRC Press, Taylor and Francis Group; 2019

[19] Polk C. Introduction: III DC and low frequency magnetic fields. In: Polk C, Postow E, editors. Handbook of Biological Effects of Electromagnetic Fields. 2nd ed. Boca Raton: CRC Press; 1996. p. 7

[20] Gray JR, Frith CH, Parker JD. In vivo enhancement of chemotherapy with static electric or magnetic fields. Bioelectromagnetics. 2000;**21**:575-583

[21] Deno DW, Carpenter DO. I.1. Sources and characteristics of electric and magnetic fields. In: Carpenter DO, Ayrapetyan SN, editors. Biological Effects of Electric and Magnetic Fields. Vol. 1. San Diego: Academic Press; 1994. pp. 18-19

[22] USGS. Available from: https://www.usgs.gov/faqs/how-does-earths-core-generate-a-magnetic-field#:~:text=The%20Earth%27s%20outer%20core%20is,to%20electrical%20and%20magnetic%20energy

[23] University of North Carolina. Available from: https://oceanweb.sites.oasis.unc.edu/www.unc.edu/depts/oceanweb/turtles/geomag.html

[24] Martino CF, Portelli L, McCabe K, et al. Reduction of the Earth's magnetic field inhibits growth rates of model cancer cell lines. Bioelectromagnetics. 2010;**31**(8):649-655

[25] Roth BJ. Biomagnetism: The first sixty years. Sensors. 2023;**23**:4218. DOI: 10.3390/s23094218

[26] Andersson LC, Nilsson K, Gahmberg CG. K562--a human erythroleukemic cell line. International Journal of Cancer. 1979;**23**(2):143-147. DOI: 10.1002/ijc.2910230202

[27] Cellosaurus - An Online Encyclopedia of Cell Lines. Available from: https://www.cellosaurus.org/CVCL_6868

[28] Cellosaurus - An Online Encyclopedia of Cell Lines. Available from: https://www.cellosaurus.org/CVCL_6890

[29] Uribe ML, Marocco I, Yarden Y. EGFR in cancer: Signaling mechanisms, drugs, and acquired resistance. Cancers (Basel). 2021;**13**(11):2748. DOI: 10.3390/cancers13112748

[30] Fan Z, Hu P, Xiang L, et al. A static magnetic field inhibits the migration and telomerase function of mouse breast cancer cells. BioMed Research International. 2020;**2020**:7472618. DOI: 10.1155/2020/7472618

[31] Zhang L, Wang J, Wang HL, et al. Moderate and strong static magnetic fields directly affect EGFR kinase domain orientation to inhibit cancer cell proliferation. Oncotarget. 2016;**7**(27):41527-41539

[32] Song C, Yu B, Wang J, et al. Moderate static magnetic fields suppress ovarian cancer metastasis via ROS-mediated oxidative stress. Oxidative Medicine and Cellular Longevity. 2021;**2021**:7103345. DOI: 10.1155/2021/7103345

[33] Luo Y, Ji X, Liu J, et al. Moderate intensity static magnetic fields affect mitotic spindles and increase the antitumor efficacy of 5-FU and Taxol. Bioelectrochemistry. 2016;**109**:31-40

[34] Liu Y, Qi H, Sun R-G, et al. An investigation into the combined effect of static magnetic fields and different anticancer drugs on K562 cell membranes. Tumori. 2011;**97**:386-392

[35] Jalali A, Zafari JFJ, et al. Combination of static magnetic field and cisplatin in order to reduce drug resistance in cancer cell lines. International Journal of Radiation Biology. 2019;**95**(8):1194-1201

[36] Zafari J, Rastegar-Pouyani N, Jouni FJ, et al. Static magnetic field reduces cisplatin resistance via increasing apoptosis pathways and genotoxicity in cancer cell lines. Scientific Reports. 2024;**14**:5792. DOI: 10.1038/s41598-024-56605-1

[37] Chen W-F, Qi H, Sun R-G, et al. Static magnetic fields enhanced the potency of cisplatin on K562 cells. Cancer Biotherapy and Radiopharmaceuticals. 2010;**25**(4):401-408. DOI: 10.1089/cbr.2009.0743

[38] Kamalipooya S, Abdolmalecki P, Salemi Z, et al. Simultaneous application of cisplatin and static magnetic field enhances oxidative stress in HeLa cell line. In Vitro Cellular & Developmental Biology - Animal. 2017;**53**:783-790

[39] Verdom B, Abdolmaleki P, Behmanesh M. The static magnetic field remotely boosts the efficiency of doxorubicin through modulating ROS behaviors. Scientific Reports. 2018;**8**:990. DOI: 10.1038/s41598-018-19247-8

[40] Gellrich D, Becker S, Strieth S. Static magnetic fields increase tumor microvessel leakiness and improve antitumoral efficacy in combination with paclitaxel. Cancer Letters. 2014;**343**:107-114

[41] Vergallo C, Ahmadi M, Mobasheri H, et al. Impact of inhomogeneous static magnetic field (31.7 – 232.0 mT) exposure on human Neuroblastoma SH-SY5Y cells during cisplatin administration. PLoS One. 2014;**9**(11):e113530. DOI: 10.1371/journal.pone.0113530

[42] Yamamoto T, Nagayama Y, Tamura M. A blood-oxygentation-dependent increase in blood viscosity due to a static magnetic field. Physics in Medicine and Biology. 2004;**49**(14):3267-3277. DOI: 10.1088/0031-9155/49/017

[43] Sud R, Khorana AA. Cancer-associated thrombosis: Risk factors, candidate biomarkers and a risk model. Thrombosis Research. 2009;**123**(Suppl 4):S18-S21. DOI: 10.1016/S0049-3848(09)70137-9

[44] Sun C, Yu H, Wang X, et al. A pilot study of extremely low-frequency magnetic fields in advanced non-small cell cancer: Effects on survival and palliation of general symptoms. Oncology Letters. 2012;**4**(5):1130-1134. DOI: 10.3892/OL.2012.867

[45] Ronchetto F, Barone D, Cintorino M, et al. Extremely low frequency-modulated static magnetic fields to treat cancer: A pilot study on patients with advanced neoplasm to assess safety and acute toxicity. Bioelectromagnetics. 2004;**25**(8):563-571. DOI: 10.1002/bem.20029

[46] Zhu M, Yang Z, Yu H, et al. The efficacy and safety of low-frequency rotating static magnetic field therapy combined with chemotherapy on advanced lung cancer patients: A randomized,

double-blinded, controlled clinical trial. International Journal of Radiation Biology. 2020;**96**(7):943-950. DOI: 10.1080/09553002.2020.1748737

[47] Salvatore JR, Harrington J, Kummet T. Phase I clinical study of a static magnetic field combined with anti-neoplastic chemotherapy in the treatment of human malignancy: Initial safety and toxicity data. Bioelectromagnetics. 2003;**24**:524-527

[48] Bailey A, Diven M, Salvatore J. The use of magnets, magnetic fields, and copper devices in a veteran population. Federal Practitioner. 2023;**40**(Suppl 3): S106-S110a. DOI: 10.12788/fp.0397

## Chapter 9

# High-Performance Magnetic-Bottle Electron Energy Spectrometer Based on Side Pole-Piece Magnetic Lens

*Yifan Kang, Hongtao Yang, Chao Wang and Wei Zhao*


## Abstract

The photoelectron energy spectrometer is known as "fingerprint technology" in atomic and molecular physics, surface science, and other research fields, the core of which is the precise spatiotemporal control of electron packets. Because the scientific issues involved, such as the space charge effect of pulsed electron packets, have not been clearly elucidated, the spectrometer technology with high energy resolution, wide detection range, and high collection efficiency is still unavailable. A novel physical model for analyzing the space charge effect of electron packets, the reduced thin electron disk model (RETED), is proposed. It is found that the initial parameters of the electron packet have complicated coupling effects on its broadening magnitude and that the self-broadening process can be decomposed into an avalanche explosion process and a subsequent quasi-linear spreading. The avalanche explosion occurs within an extremely localized time and space range and is identified as the dominating factor for the electron packet broadening process. To further aim at constructing a set of adiabatic magnetic field architectures suitable for a magnetic-bottle time-of-flight electron energy spectrometer, a practical scheme for a side pole-piece magnetic lens is proposed, and the high peak-valley ratio adiabatic magnetic field is constructed by combining a side pole-piece magnetic lens, a pole-piece permanent magnet, and a solenoid. Finally, a high-performance magnetic-bottle electron energy spectrometer is demonstrated. The maximum detectable electron energy can reach 1000 eV, the full-range relative energy resolution is better than 1%, and the collection solid angle reaches $2\pi$. These results will enrich the theory of high-precision electron energy spectra detection techniques, meeting the urgent demand in such fields as electron momentum spectroscopy (EMS) and photoelectron spectroscopy.


**Keywords:** space charge effect, electron energy spectrometer, adiabatic magnetic field, side pole-piece magnetic lens, electronic trajectory tracking

## 1. Introduction

The exploration of the essence of matter has now delved into the microscopic subatomic field [1, 2], using photons and/or electrons as the main probe medium. Electron energy/momentum spectroscopy detection using photons as primary probe particles is often referred to as the photoelectron spectroscopy technique, which is an extremely important research branch in the field of spectroscopy technology. Currently, with the rapid expansion of research depth and breadth, exploring new high-precision detection and analysis techniques for electron energy/momentum spectra has become a common research hotspot in fields of attosecond spectroscopy [3–5] (1 attosecond = $10^{-18}$ seconds), electron momentum spectroscopy (EMS) [6, 7], and photoelectron spectroscopy [8, 9].

The emergence of attosecond spectroscopy, which directly studies subatomic electron dynamics, is a revolutionary breakthrough in fundamental sciences in the twenty-first century. As the core medium of spectroscopy, both the characterization of attosecond laser pulse width and the spectroscopic research mostly involve the detection of photoelectron energy/momentum spectra of target atom/molecule [10, 11], and the spectrum is the core basis for the inversion of subatomic electron dynamics information to be diagnosed. Therefore, the electron spectrometer is not only an indispensable device for attosecond spectroscopy, but its detection performance is also crucial and determines the measurement accuracy of subatomic electron dynamics, thereby determining the application scope of attosecond spectroscopy. Therefore, the fine detection technique of electron spectrum is known as the "fingerprint technology" in attosecond spectroscopy [1, 11]. As for attosecond spectroscopy, the photoelectrons have a broad energy spectrum width almost equal to that of attosecond lasers. To guarantee a decent signal-to-noise ratio, the photoelectron detection is required to simultaneously possess the performance of a wide detection range, high collection efficiency, and high energy resolution [1, 4, 11–16]. Meanwhile, differential pumping is often used to meet the vacuum requirements [5, 11–16], which often becomes an important factor limiting the electron collection efficiency and detection range of the system. For example, in engineering applications, it often limits the electron collection angle to within 60 degrees, which results in an electron collection efficiency of only below 10% [17], assuming that electron emission is isotropic. The international level of laboratory-based attosecond laser technology is as follows: the shortest packet width reaches 43 as (spectral full width of about 100 eV) [13, 14], and the highest photon energy reaches 250 eV [17, 18]. According to the application requirements of laser spectroscopy, generating soft X-ray isolated attosecond lasers with a smaller packet width and higher photon energy is one of the research hotspots [18–20]. For a Fourier transform-limited laser packet of one atomic time unit ($\sim$25 as), the corresponding photoelectron spectrum full width at half maximum is about 75 eV, and the full energy spectrum span will reach about 200 eV. For such a wide-spectrum attosecond laser probe, the requirements for photoelectron spectrum detection will be higher. The common technical solution is to tune the electrons' energy for segmented detection, then to perform energy spectrum splicing to complete a full spectrum measurement. This will otherwise introduce other measurement errors, such as light source jitter and changes in experimental conditions. Therefore, exploring new ideas to improve the detection range and collection efficiency of electronic detection systems has become a hot topic of continuous attention in this field in recent years [1, 4].

EMS can accurately obtain the energy spectrum and wave function information of electron orbitals in atomic molecules through the coincidence measurement of scattered and ionized electrons. It is currently an advanced experimental method for studying the microscopic properties of atoms and molecules, solid surfaces, and thin films. The electron momentum spectrometer is the most critical technical equipment among them [4, 21, 22]. The continuous expansion of EMS research over the past years is closely related to the updating and development of electronic momentum spectrometer technology. Weigold et al. successfully developed the first-generation (e, 2e) electron momentum spectrometer in 1973 [23], which used a cylindrical analyzer as an energy-dispersive electron-optical system and could only perform single-angle and single-energy point coincidence counting. Afterward, with the introduction of new electron dispersion systems and position-sensitive detectors, second-generation spectrometers [24] (using hemispherical energy analyzers or cylindrical analyzers to generate energy dispersion with an energy resolution of about 1.5 eV) and third-generation spectrometer technology [25, 26] (using drum-type energy analyzers as energy dispersion systems with a detection angle utilization rate of up to about 25% and an energy resolution of 0.5–1.5 eV) emerged successively, which can achieve multichannel measurements. Nowadays, the fourth-generation spectrometers have been achieved [27], which have been applied to study such phenomena as molecular electron correlation effects and vibration effects. Although it has the advantages of high detection efficiency and a relatively wide momentum detection range, the dilemma still exists that the maximum detectable energy is still relatively low, currently mostly in the range of tens of electron volts, and the full-range energy resolution is still relatively low, currently around 0.5 eV. As EMS progresses deeper, the research object will be more refined physical phenomena related to the energy level structure of core electrons, such as the Jahn-Teller [28] and spin-orbit splitting effect [29], which requires the spectrometer to have even higher comprehensive performance [3, 6, 29]. Currently, synchrotron radiation sources have been an important comprehensive research platform for cutting-edge innovative research [30, 31], involving the diagnostics of emitted photoelectrons and secondary ions from the interactions of synchrotron radiation with matter [32, 33]. The synchrotron radiation photoelectron spectroscopy, along with customized experimental methods such as the time-resolved diagnostic technique, greatly expands the breadth and depth of photoelectron spectroscopy. With the continuous upgrades of the fourth-generation synchrotron radiation light source of free electron laser [34, 35], soft/hard X-ray lasers with shorter packet width and higher photon energy have rapidly entered the fields of physics, chemistry, and other basic and applied sciences [36, 37]. Consequently, the exploration of wide-spectrum high-resolution detection technology for high-energy electrons has become a research focus [38].

Particle momentum is the expression of energy in spatial dimension, so the essence of electron momentum spectrum detection is energy spectrum detection [2, 6, 27]. The core of the photoelectron energy spectrometer is the energy spectrum dispersion mechanism and dispersion quality, which directly determine the main spectrometer performance, such as the energy spectrum detection range, collection efficiency, and relative energy resolution. According to the energy spectrum dispersion mechanism, current charged particle energy spectrum detection techniques can be divided into two categories: field-free [3, 4] and field-containing [11–16, 39–41]. In order to obtain a high-performance photoelectron energy spectrometer, although a lot of work has been done so far, the involved bottleneck still exists: there is a significant negative correlation suppression effect between detection range, energy resolution, and collection efficiency. This is determined by its working principle. Taking the commonly used truncated cone photoelectron time-of-flight spectrometer as an example [17], the electrons with different initial energies and emission angles have traveled the same effective axial distance before being received by the detector. Therefore, electrons with smaller axial energy components have greater time-of-flight dispersion, which means that the electro-optical system has relatively higher energy resolution for electrons with larger initial emission angles. In practical engineering applications, electrons with smaller initial emission angles have a higher probability of being detected. Meanwhile, the energy resolution performance of the system sharply decreases and deteriorates with the increase of initial energy. For time-of-flight-based energy analysis techniques, the energy dispersion can be decomposed into two categories: angular dispersion of electrons with the same initial emission energy and energy dispersion of electrons with the same initial emission angle. The former is positive dispersion: the larger the initial angle of particles, the greater the dispersion of flight time. While the latter is negative dispersion, the larger the initial energy of the particle, the smaller the dispersion during flight time. As for the existing energy spectrometers, these two types of dispersion mechanisms are coupled with each other during the entire life cycle of electrons and increase sharply with the evolution of the transit process. This makes the dispersion of flight time-related parameters very small, the dispersion quality is extremely low, and the overall performance improvement is constrained.

From the perspective of technological essence, the core of the photoelectron energy spectrometer is the precise spatiotemporal control of the pulsed electron beam. The scientific issues, such as the space charge effect of the pulsed electron beam, have not been clearly elucidated, and the spectrometer technology with high energy resolution, wide detection range, and high collection efficiency is still unavailable. The space charge effect is a tricky scientific problem in the field of charged particle optics, which originates from the inherent Coulomb repulsion between charged particles and can lead to complex changes of momentum states of pulsed electron packets. For photoelectron energy spectrometer, the space charge effect is the main factor affecting the time-resolving performance. The original momentum state of photoelectrons generated by the ionization process of atoms or molecules is called the *intrinsic momentum state*, which carries the intrinsic state information of the atomic or molecular dynamics to be diagnosed. After being modulated by the space charge effect, the intrinsic momentum state of photoelectrons becomes the *initial momentum state*. The initial momentum state of electrons directly determines their mapping trajectory and kinematic characteristics in the momentum spectrum detection system. Therefore, the mechanism of the space charge effect must be elucidated.

In this chapter, we propose a novel physical model for analyzing the self-broadening dynamics of electron packets, the reduced thin electron disk model. To further aim at the construction of an adiabatic magnetic field used by a magnetic-bottle time-of-flight electron energy spectrometer, a practical scheme for a side pole-piece magnetic lens is proposed, and the high peak-valley ratio adiabatic magnetic field is tailored by combining a side pole-piece magnetic lens, a pole-piece permanent magnet, and a solenoid. Finally, we have demonstrated a design scheme for a magnetic-bottle electron energy spectrometer. The maximum detectable electron energy can reach 1000 eV, the full-range relative energy resolution is better than 1%, and the collection solid angle reaches $2\pi$. These research results will not only enrich the theory of high-precision detection and analysis techniques for electron energy spectra, but they can also serve as a direct reference for other charged particle analysis spectrometer techniques.

## 2. Space charge effect of pulsed electron beam

For the magnetic-bottle-type spectrometer in the following section, the space charge effect of the pulsed electron beam is mainly limited in space to its initial generation position, since electrons quickly disperse along different magnetic field lines in space as the guiding magnetic field rapidly weakens. Because the guiding magnetic field is bottle-shaped, the bottleneck longitudinal magnetic field where the electron source is generated is as high as over 1 Tesla, and the Lorentz force provides strong lateral confinement for electrons. Therefore, in the following analysis, we omit the transverse expansion of the electron packet and only consider the longitudinal broadening, since the former is proved to be both small and slow by other investigations even without lateral confinement from a magnetic field [42, 43]. Based on the characteristics of many-body interactions among electrons [42], a novel physical model of a "reduced thin electron disk" is established for analyzing the self-broadening dynamics of a pulsed electron beam, in which it is differentiated into many transverse disks with thickness *dz'*, as shown in **Figure 1**. The parameter $r_0$ is the radius of the electron packet, $l(t)$ the longitudinal length at time $t$. So one can get the local built-in electric field inside the pulsed electron beam

$$
\begin{aligned}
E_{local}(z,\ t) &= \frac{\rho(t)}{2\varepsilon_v}\int_{-l(t)/2}^{z}\left[\frac{z-z'}{|z-z'|}-\frac{z-z'}{\sqrt{r_0^2+(z-z')^2}}\right]dz' \\
&\quad -\frac{\rho(t)}{2\varepsilon_v}\int_{z}^{l(t)/2}\left[\frac{z'-z}{|z'-z|}-\frac{z'-z}{\sqrt{r_0^2+(z-z')^2}}\right]dz' \\
&= \frac{\rho(t)}{2\varepsilon_v}\left[2z+\sqrt{r_0^2+\left(z-\frac{l(t)}{2}\right)^2}-\sqrt{r_0^2+\left(z+\frac{l(t)}{2}\right)^2}\right]
\end{aligned}
\tag{1}
$$

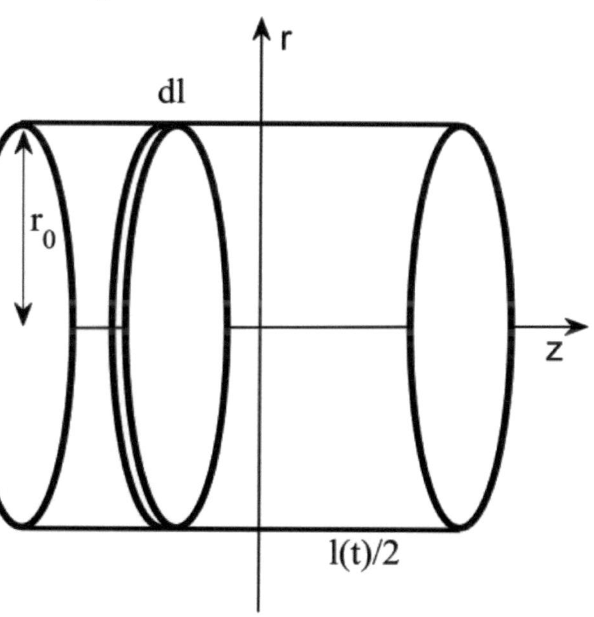


**Figure 1.**
*Analyzing the model for electron packet broadening.*

where $\rho(t) = -Ne/[\pi r_0{}^2 l(t)]$ is the space charge volume density, $N$ is the number of electrons contained in the pulsed electron beam, $e$ is the electron charge, and $\varepsilon_v$ is the free space permittivity. The electric field distribution on the center axis can be obtained, as shown in **Figure 2**. It shows that the electric field strength at the center of the electron beam is 0, while the maximum electric field strength is located at both ends of the electron beam. When the charge volume density increases, the electric field strength will increase accordingly. According to the multiparameter dependence of electric field strength, the description of the broadening dynamics of the electron beam will undoubtedly be very complicated.

As for the general case with an external electric field $E_0$, which is assumed to be along the longitudinal direction, the superposition of $E_0$ and $E_{local}$ will determine the broadening process. For typical time-of-flight photoelectron spectrometer systems, the introduction of an axial magnetic field is often used to constrain electrons laterally to improve the electron collection efficiency, and it does not affect the time-of-flight characteristics of electron transition in the axial direction. Therefore, as an example, we here consider the application scenarios with only an acceleration electric field zone and an electric-field-free drifting zone, which is shown in **Figure 3**. The region with an axial distance of $d_a$ is the accelerating zone, while the other is the drifting zone. The pulsed electron beam is generated at the electrode with applied voltage $U_p = 0$. The acceleration voltage applied to a mesh electrode is $U_a$. The drifting zone is formed between the mesh electrode and another electrode with applied voltage $U_t = U_a$. For the sake of simplicity, the pulsed electron beam is assumed to be of square shape, with an initial energy spread of $\Delta\varepsilon_0$ and a beam width of $\tau_0$. So the initial longitudinal length can be calculated as

$$l_0 = \frac{eU_a}{2m_e d_a}\tau_0{}^2 + \tau_0\sqrt{\frac{2\Delta\varepsilon_0}{m_e}}, \tag{2}$$

According to Newton's second law, one can get

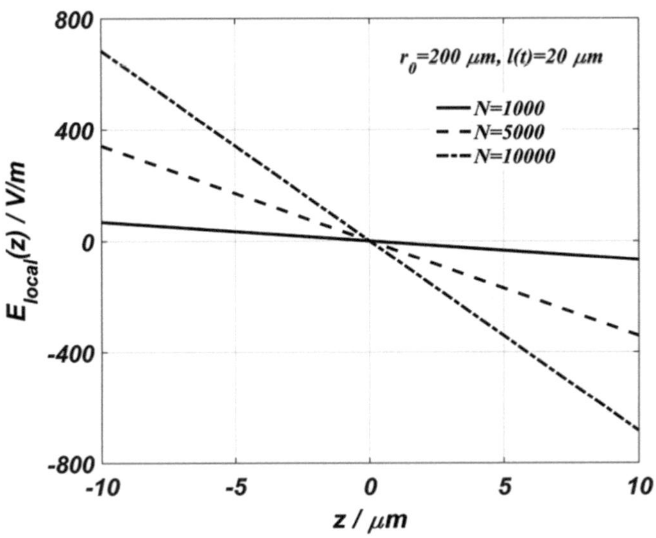


**Figure 2.**
*On-axis electric field distribution in electron packet.*

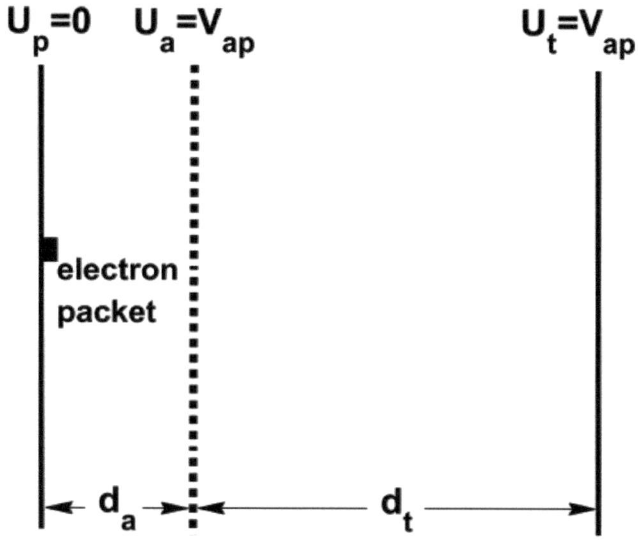


**Figure 3.**
*Analysis scenario for electron packet.*

$$\frac{d^2l(t)}{dt^2} = \frac{Ne^2}{m_e \varepsilon_v \pi r_0^2} \left(1 - \frac{l(t)}{\sqrt{l(t)^2 + 4r_0^2}}\right) \tag{3}$$

Eq. (3) serves to investigate the broadening dynamics of the electron beam.

The broadening evolution characteristics of the pulsed electron beam in the acceleration region are analyzed quantitatively, and the results are shown in **Figure 4**. When N is chosen as the only variable parameter in **Figure 4(a)**, the generated 200 fs electron beam rapidly widens to 310 fs within a very short travel distance shown in **Figure 4(d)** and then undergoes an approximately linear broadening process, with the broadening rate being related to parameter N. As for the case when $r_0$ is the only variable parameter in **Figure 4(c)**, one can see that from a qualitative perspective, the dynamic characteristics of broadening are almost the same. It shows that the broadening process consists of two sequential steps of early happening avalanche explosion and subsequent quasi-linear spreading. For the former, it occurs within an extremely localized range of less than 1 micrometer and serves as the dominating factor for the broadening process. It should be noted that, as shown in **Figure 4(b)**, increasing the electric field strength in the acceleration region helps to suppress the space charge effect.

The coupling effects of initial parameters on the quasi-linear spreading process are studied. When there are two variable parameters of N and $r_0$, as in **Figure 5**, the dependence of the broadening process on parameters will be very complex. After detailed analysis, we found that, for a given initial packet width $\tau_0$, the magnitude of self-broadening is approximately proportional to the combined parameters of $N^{1.1}/r_0^2$. Furthermore, when the parameters of $\tau_0$, $r_0$, and $N$ all are free to change, the situation becomes even more complicated, as in **Figure 6,** and a quantitative description of broadening characteristics can only be given through calculations.

For the magnetic-bottle electron spectrometer to be discussed in the next section, the external electric field where the pulsed electron beam is born is zero; thus, one

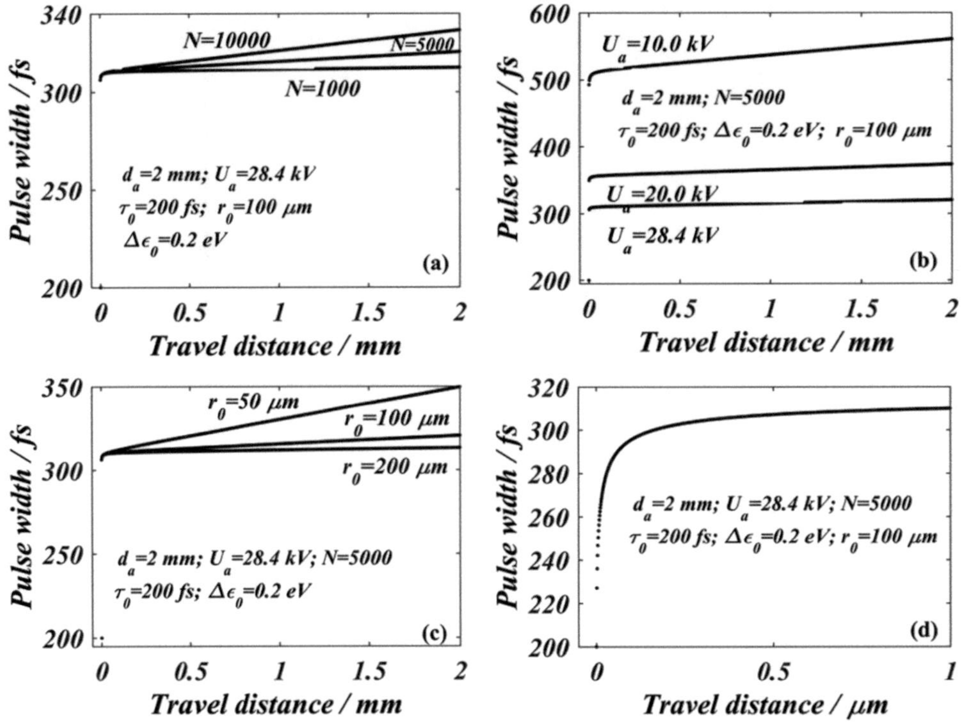


**Figure 4.**
*Broadening characteristics of electron packet in accelerating field: (a) N; (b) $U_a$; (c) $r_0$; and (d) avalanche broadening.*

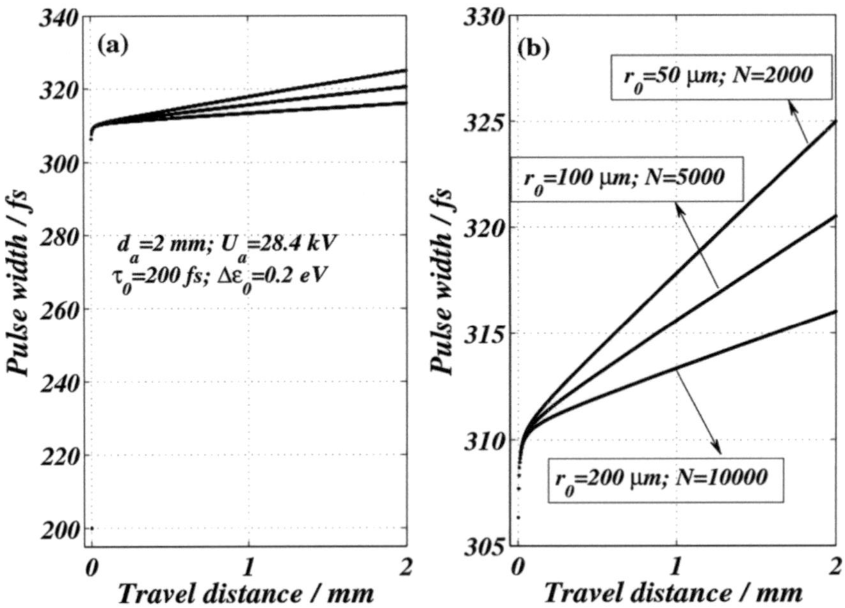


**Figure 5.**
*Coupled effect of $r_0$ and N on the quasi-linear spreading process.*

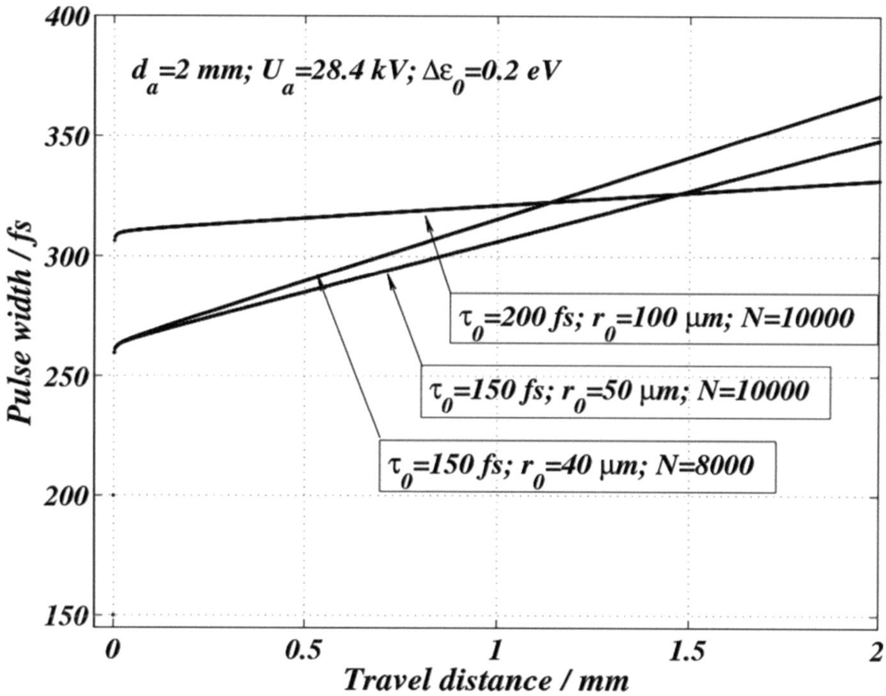


**Figure 6.**
*Coupled effect of $\tau_0$, $r_0$, and N on the quasi-linear spreading process.*

actually deals with the space charge effect in an electric-field-free drifting zone. For such an electron-optical device, the space charge effect is one of the main factors affecting the time- or energy-resolving performance. The original photoelectron's momentum distribution determined by the photoionization process is called the intrinsic momentum state of the electron beam. Once the electron is generated, the space charge effect takes effect immediately. After being modulated, the intrinsic momentum state of photoelectrons becomes the initial momentum state, which directly determines their trajectory characteristics in the momentum spectrum detection system. Therefore, the quantitative calculation of the space charge effect will ensure the accuracy of the design of the magnetic-bottle electron spectrometer.

## 3. Magnetic-bottle electron energy spectrometer

### 3.1 Fundamentals for magnetic-bottle spectrometer

The technique of the magnetic-bottle spectrometer was invented in 1983 by Kruit and Read [44], as shown schematically in **Figure 7**. It naturally has superior comprehensive performance of high collection efficiency and satisfactory energy resolution in a decent detection range. For an electron initially emitted at an angle $\theta_i$ to the spectrometer axis, with a velocity $v_i$, it undergoes helical motion in the field $B_i$. The Lorentz force causes it to spiral around a magnetic field line, as depicted in **Figure 7**. The angular frequency (cyclotron frequency) of the motion is

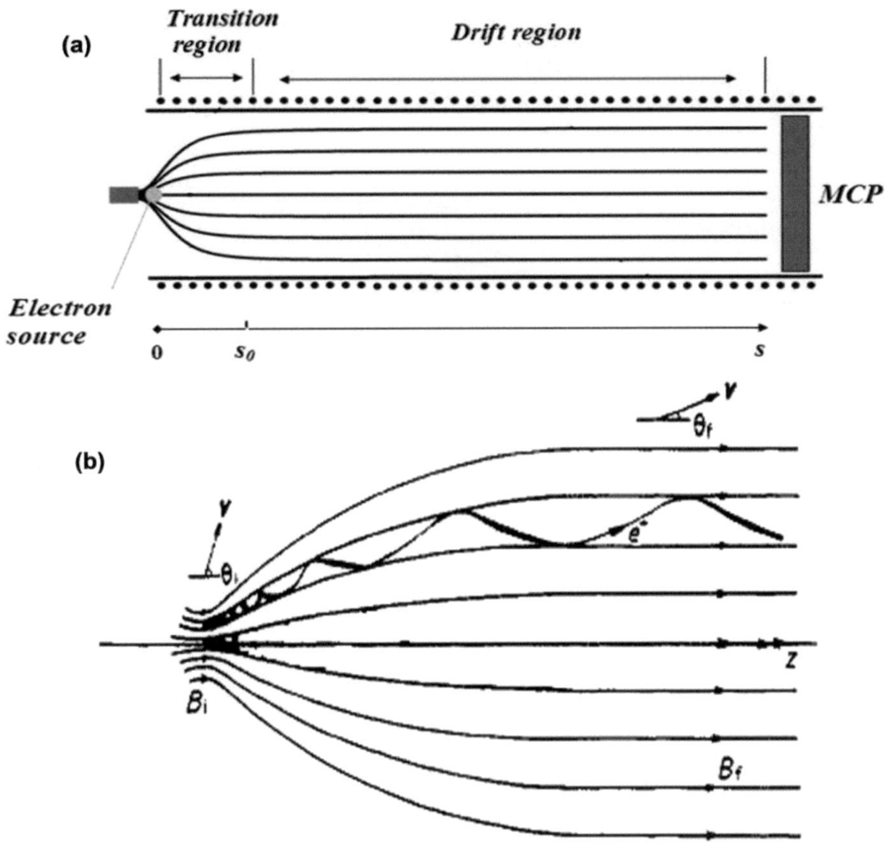


**Figure 7.**
*(a) A magnetic-bottle electron spectrometer; (b) the helical motion of an electron moving in a magnetic field that changes gradually from a strong field $B_i$ to a weaker uniform field $B_f$ [44].*

$$\omega_i = \frac{e}{m_e} B_i \tag{4}$$

where $e$ and $m_e$ are the charge and mass of an electron, respectively. The radius of the orbit (cyclotron radius) of an electron with initial speed $v_i$ is

$$r_i = \frac{v_i \sin \Theta_i}{\omega_i}. \tag{5}$$

The angular momentum of the circular motion is

$$\ell_i = \frac{m_e^2 v_i^2 \sin^2 \Theta_i}{eB_i}. \tag{6}$$

Then the electron enters the region of the weak uniform magnetic field $B_f$. If the field experienced by an electron changes negligibly in the course of one revolution of the helical motion, the change in magnetic field with z will be considered adiabatic. Under these conditions, the angular momentum is a conserved quantity.

To quantitatively describe this condition of adiabatic approximation, an adiabaticity parameter $\chi_1$ is introduced. For an electron orbiting around the z-axis at an angle $\theta_i$ to the axis, the pitch magnitude, i.e., the longitudinal distance that the electron moves during one orbit when the electron is near the position z, is

$$\Delta z_{pitch} = \frac{2\pi}{\omega} v_z = 2\pi \frac{m_e}{e} \frac{v_i \cos\theta_i}{B_z}, \tag{7}$$

and the change in $B_z$ over this distance is

$$\Delta B_z = \frac{dB_z}{dz} \Delta z_{pitch}. \tag{8}$$

The adiabaticity parameter is defined by

$$\chi_1 = \left. \frac{\Delta B_z}{B_z} \right|_{\theta \to 0} = 2\pi \frac{m_e}{e} \frac{v_i}{B_z^{\ 2}} \left| \frac{dB_z}{dz} \right|, \tag{9}$$

which is the fractional change in the field that the electron experiences in one cyclotron orbit. The value of $\chi_1$ should be less than 1 for 1 eV electrons to ensure that the electron motion is adiabatic.

Under the adiabatic condition, the inhomogeneous magnetic-bottle magnetic field effect will result in the parallelization of the electrons' trajectories. Assuming $B_i = 1\,T$ and $B_f = 1\,mT$, the angle of the helical motion in the region of low field is given by

$$\frac{\sin^2\Theta_f}{\sin^2\Theta_i} = \frac{B_f}{B_i} \approx 10^{-3} \tag{10}$$

The transverse component of the velocity is thus reduced. Since the total velocity is unchanged, the longitudinal component increases from $v_i \cos\theta_i$ to

$$v_{zf} = v\sqrt{1 - \frac{B_f}{B_i}\sin^2\Theta_i} \tag{11}$$

The time-of-flight in the drift tube is

$$t = \frac{1}{v_i}\int_0^{s_0} \frac{dz}{\sqrt{1 - \frac{B_z}{B_i}\sin^2\theta_i}} + \frac{(s - s_0)}{v_i}\left[1 + \frac{1}{2}\frac{B_f}{B_i}\sin^2(\Theta_i)\right] \tag{12}$$

The first item on the right side represents the dispersion dependence of flight time on its initial emission angle, labeled $D_{angle}$, while the second one is on its initial energy, labeled $D_{energy}$ . To possess a high electron energy resolution, one needs to decrease the absolute value of the former and increase the latter. To achieve the best energy resolution, it is important that the changing-field transition region should be as short as possible ($\left|\frac{dB_z}{dz}\right|$ as large as possible) while remaining in the adiabatic limit. These two requirements are contradictory, and a compromise must be found.

### 3.2 Side pole-piece magnetic lens

In order to enhance the magnetic field strength at the electron source, a short magnetic lens is often used. As for a conventional double-pole-piece magnetic lens, there is an axial pole gap and two pole faces, which determine the axial magnetic flux density distribution and lens properties. Whether the geometry is symmetrical or asymmetrical, the peak axial flux density is always located within the lens body. The feature is illustrated qualitatively in **Figure 8(a)**. Here, a new unconventional magnetic lens with a side pole gap is investigated, as shown in **Figure 8(b)**. The air gap is radially annular and open at one side of the lens case.

In case that there is no saturation in the magnetic circuit, i.e., the magnetic material is assumed to be linear, the scalar magnetic potential has been used for calculation, and the surfaces of two side pole pieces are of equipotential with magnetic potential difference equaling the lens excitation *NI*,

$$U_b - U_a = NI, \tag{13}$$

where $U_b$ and $U_a$ denote magnetic scalar potentials of the outer and inner pole pieces, respectively, as shown in **Figure 9**.

In case there is no inner bore (i.e., $r_c \to 0$), Green's theorem is used to express the magnetic potential for the region outside the lens case ($z < 0$ half-space):

$$\phi = \frac{-1}{4\pi}\iint \phi \frac{\partial G}{\partial n} dA' = \frac{-1}{4\pi}\iint \phi \frac{\partial G}{\partial n} r' dr' d\theta', \tag{14}$$

where $G$ is the Green function of the Dirichlet problem, the integral region is the plane $z = 0$, n denotes the external normal of the boundary of the half-space, and $(r, \theta, z)$ is a cylindrical coordinate system. To an acceptable accuracy, the potential distribution at the plane $z = 0$ can be assumed as (set $U_a = 0$, $U_b = NI$)

$$\phi(r', 0) = \phi(r') = \begin{cases} 0, r' \le r_a \\ NI \ln \frac{r'}{r_a} / \ln \frac{r_b}{r_a}, r_a < r' < r_b, \\ NI, r' \ge r_b \end{cases} \tag{15}$$

substituting (15) into (14) gives

$$\phi(r, z)|_{z<0} = \frac{-1}{4\pi}\int_0^{2\pi} d\theta' \int_0^{\infty} r' dr' \phi(r') \frac{\partial}{\partial z}\left[\frac{1}{R} - \frac{1}{R'}\right], \tag{16}$$

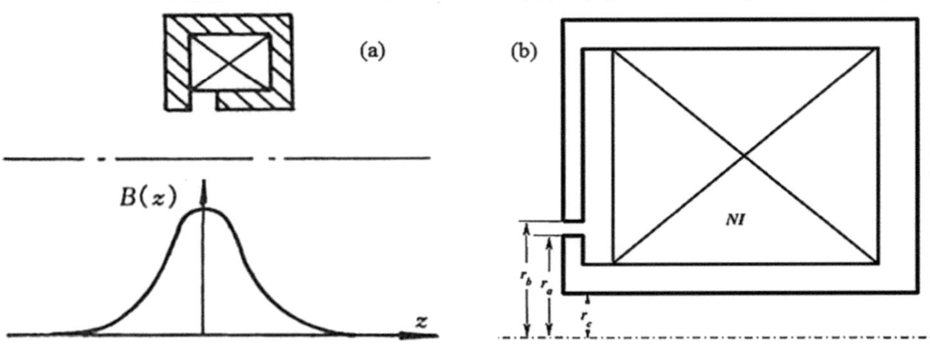


**Figure 8.**
*(a) Conventional and (b) side pole-piece magnetic lens.*

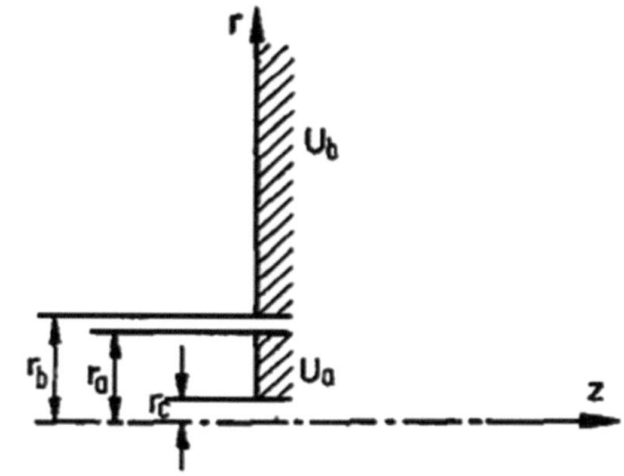


**Figure 9.**
*Analytical model of the side pole-gap magnetic electron lens.*

where the term within the square brackets is the Green function of the Dirichlet problem for the half-space, and $R$ and $R'$ denote the distance between the "source point" and the "field point" and its image specifically. For first-order Gaussian optical properties and aberration coefficient calculation, only axial field distribution $B(z)$ is needed. Taking the rotational symmetry of the field into consideration, for an axial "field point," one can get

$$B(z) = \frac{\mu_0 NI}{ln\,\frac{r_b}{r_a}}\left(\frac{1}{\sqrt{z^2 + r_a{}^2}} - \frac{1}{\sqrt{z^2 + r_b{}^2}}\right). \tag{17}$$

As for the case that $r_c$ is much smaller than $r_a$, it is reasonably assumed that the effect of the inner bore on the field distribution should be limited to the neighborhood of the bore about the range of $r_c$. The field within the inner bore and nearby can be approximated by the field of a single diaphragm. The axial flux density of a side pole-gap lens with an inner bore is found to be

$$B(z) = \frac{-\mu_0 NI}{ln\,\frac{r_b}{r_a}}\left[\frac{r_b - r_a}{2r_b r_a} + \frac{r_b - r_a}{\pi r_b r_a}\left(\frac{z\bullet r_c}{z^2 + r_c{}^2} + tg^{-1}\frac{z}{r_c}\right) + \frac{1}{\sqrt{z^2 + r_b{}^2}} - \frac{1}{\sqrt{z^2 + r_a{}^2}}\right], \text{for } z < 0, \tag{18}$$

and

$$B(z) = \frac{-\mu_0 NI}{ln\,\frac{r_b}{r_a}}\left[\frac{1}{2} - \frac{1}{\pi}\left(\frac{zr_c}{z^2 + r_c{}^2} + tg^{-1}\frac{z}{r_c}\right)\right]\bullet\frac{r_a - r_b}{r_a \bullet r_b}, \text{for } z \geq 0. \tag{19}$$

The tuning characteristics of the magnetic field distribution of the side pole-gap magnetic lens are illustrated in **Figures 10–12**. The theoretical analysis shows that the flux density has a highly asymmetrical feature, and the peak flux density is always outside the lens case ($z = 0$ suface). This feature makes it possible to adjust the spatial position of the magnetic field peak to the position of the electron source. As can be seen from the next section's discussion, these features render this kind of magnetic electron lens superior for application in magnetic-bottle electron energy spectrometers.

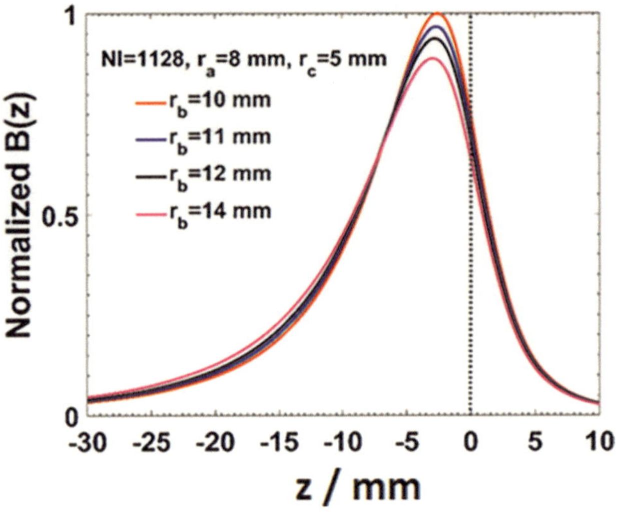


**Figure 10.**
*The dependence of the axial magnetic field on the parameter* $r_b$.

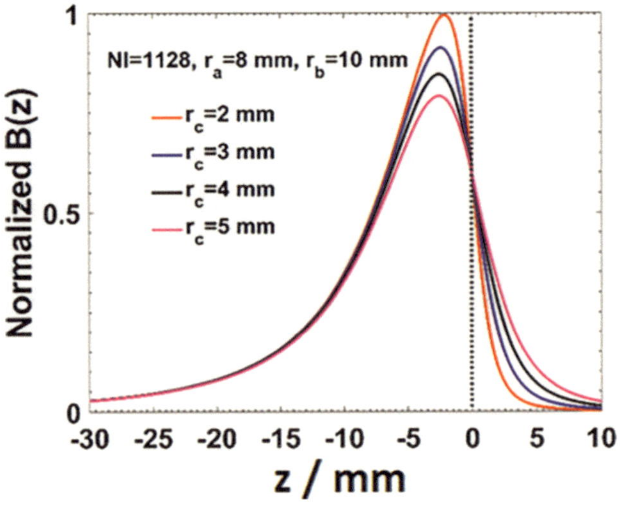


**Figure 11.**
*The dependence of the axial magnetic field on the parameter* $r_c$.

### 3.3 Magnetic-bottle electron energy spectrometer

The scheme of the proposed side pole-piece magnetic lens-based magnetic-bottle electron energy spectrometer is given in **Figure 13**. The high peak-valley ratio adiabatic magnetic field is constructed by the side pole-piece magnetic lens, the permanent magnet pole piece, and the solenoid. The strong magnetic field higher than 2 T is created by the permanent magnet with a pole piece and the side pole-piece magnetic lens. The weak field is generated by solenoid coils wrapped around the drift tube. The coils are then wrapped with double-layer μ-metal shielding to minimize the earth's magnetic field in the vacuum chamber.

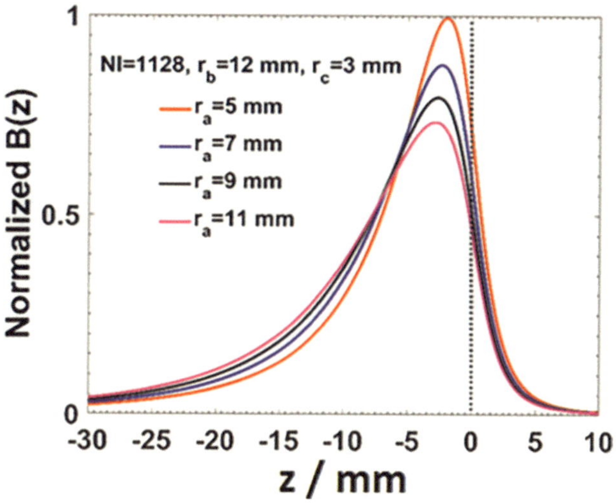


**Figure 12.**
*The dependence of the axial magnetic field on the parameter $r_a$.*

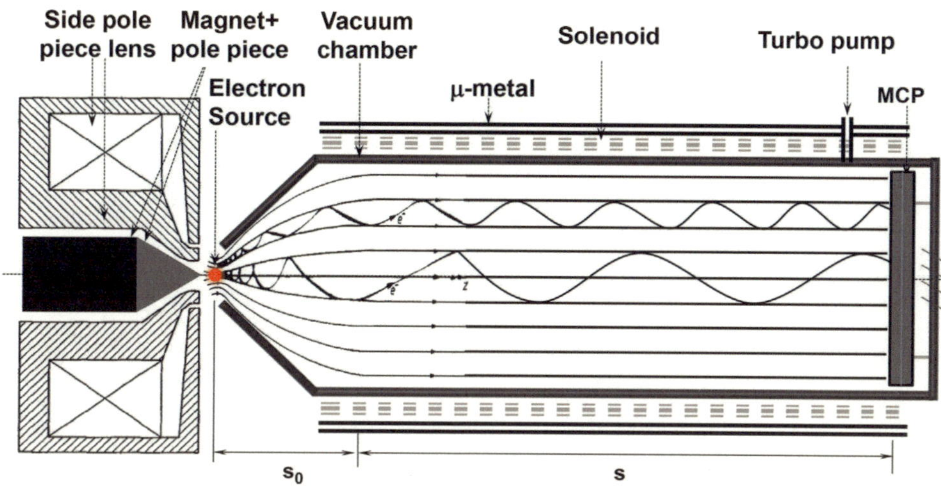


**Figure 13.**
*Schematic diagram of a magnetic-bottle electron energy spectrometer.*

The configuration design of the combined permanent magnet and pole piece is shown in **Figure 14**. The pole piece can be made of iron or permendur. Pure iron transmits a flux density of 2.1 T. Permendur is a cobalt steel with a chemical composition of 49% Co-2% V-Fe and boasts a high saturation flux density and high permeability. The maximum tolerable flux density of permendur is 2.4 T. The parameters of the cylindrical strong rare-earth Neodymium Iron Boron (NdFeB) permanent magnet are: $h = 50.8\ mm$, $r_1 = 12.7\ mm$, $M = 11000 A/m$ (magnetization). **Figure 15** shows the axial magnetic field distribution without the pole piece. One can see that the magnetic field is dropping slowly from the maximum of around 6000 Gs to around 2500 Gs at an axial position of 1.2 cm away. For the given dimensional parameters of

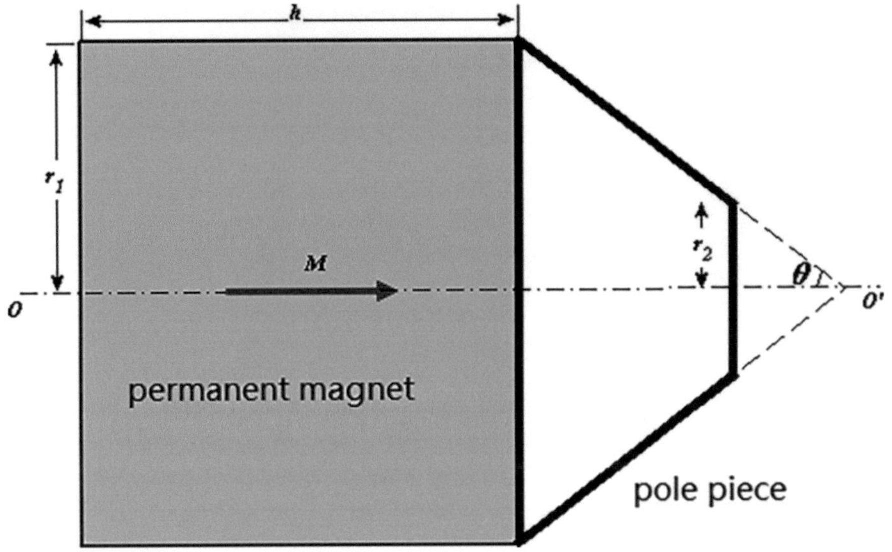


**Figure 14.**
*Combined permanent magnet and pole piece.*

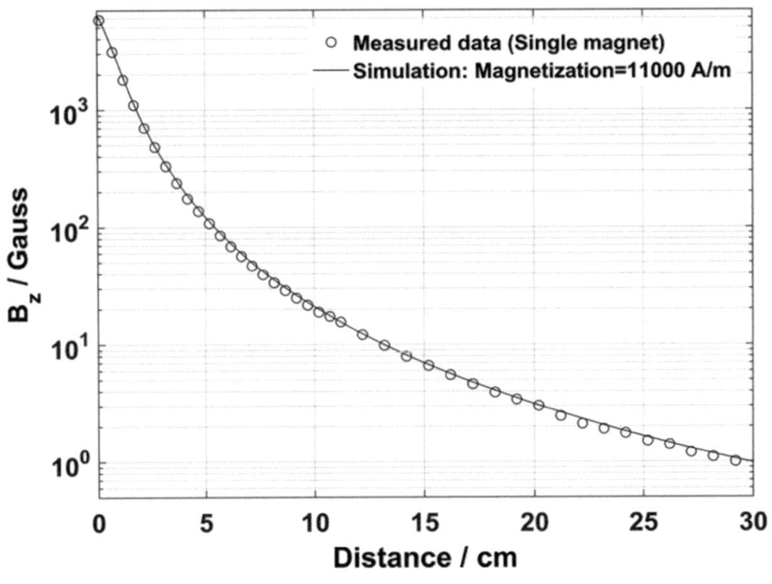


**Figure 15.**
*Magnetic field distribution along the symmetric axis of the NdFeB magnet.*

the permanent magnet, the magnetic focusing properties of the pole piece are definitely determined by $r_2$ and $\theta$. The simulations show that the peak magnetic field strength $B_{max}$ at the center of the pole piece tip can be above 1.2 Tesla when the parameter condition is satisfied, $0.8\ mm \leq r_2 \leq 1.2\ mm$ and $42^\circ \leq \theta \leq 48^\circ$, as shown in **Figure 16**. The results here state clearly the magnetic field focusing by the permendur pole piece, with the magnetic field dropping quickly from $B_{max} = 1.2\ T$ to around

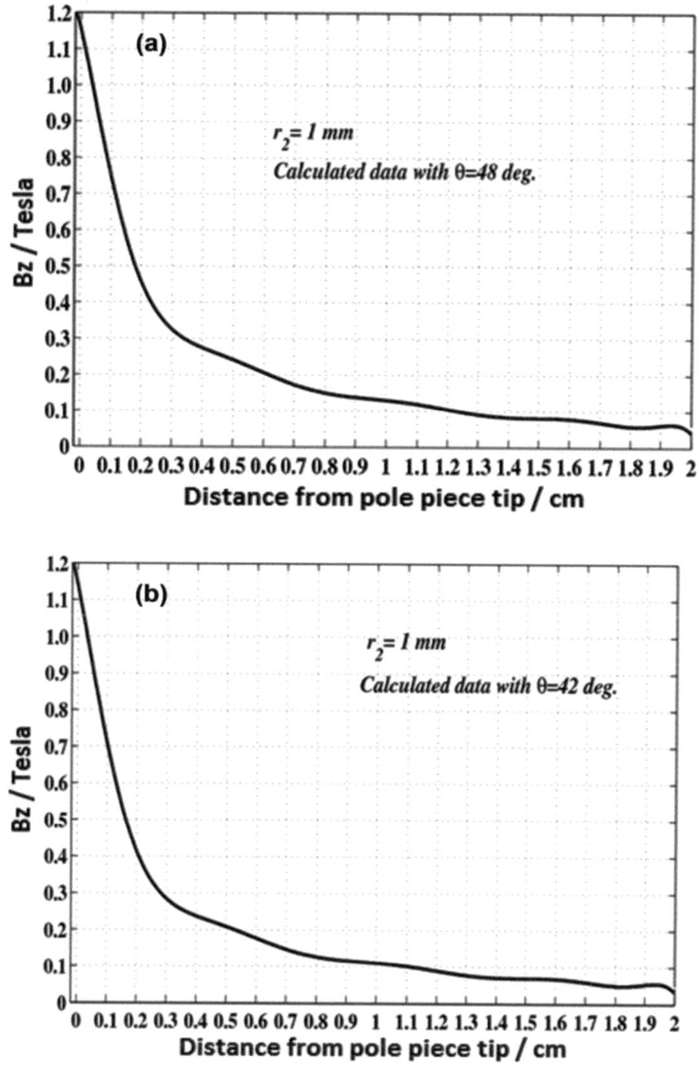


**Figure 16.**
*Magnetic field distribution along the symmetric axis of the NdFeB magnet and the permendur alloy pole piece: (a) $\theta = 48$ degs and (b) $\theta = 42$ degs.*

0.1 T at an axial position of 1.2 cm away. It is worth noting that, different from the usual intuition, a smaller size of pole piece tip will not give better focusing, with $B_{max}$ below 1.0 T. We attribute these phenomena to magnetic flux leakage due to magnetic saturation in the permendur pole piece.

The homogeneous weak magnetic field of 10 Gs is provided by a solenoid with diameter $r_s = 30\ mm$ and axial length $L_s = 1\ m$. The constructed inhomogeneous adiabatic magnetic field is given in **Figure 17**, and the peak-valley ratio of the adiabatic magnetic field is 2000:1. The electron source is located at position z = 0, and the magnetic field flux reaches up to more than 2.0 T. The electrons emitted from the electron source at different initial emission angles have been well collimated and focused, as illustrated in **Figure 18**. It shows that the collection solid

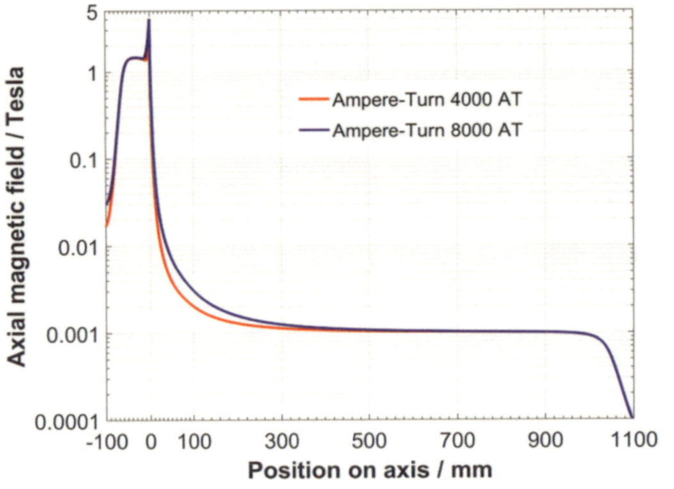


**Figure 17.**
*Axial distribution of the combined high peak-valley ratio adiabatic magnetic field.*

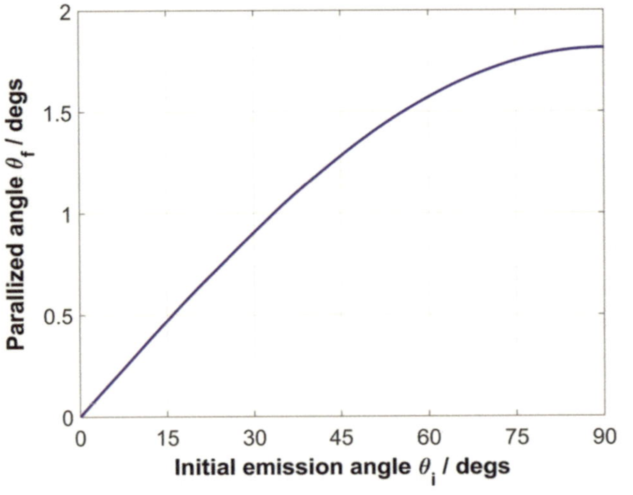


**Figure 18.**
*Parallelization focusing effect on electrons in an inhomogeneous magnetic field.*

angle can reach $2\pi$. Assuming the timing resolution $\Delta t_{min} = 1\, ns$ for the overall microchannel plate (MCP) detection system, the energy-resolving performance of the magnetic-bottle spectrometer for electrons with initial energy up to 1000 eV is calculated, and the relative energy resolution is 0.9%, as in **Figure 19**. We analyze the electron transition dynamics in the magnetic-bottle magnetic field systems, and the results are shown in **Figure 19**. Comparing **Figure 20(a)** and **(b)**, it can be seen that, on the one hand, the higher the electron energy, the larger the distribution radius of its impact on the MCP, that is, the weaker the focusing effect of the magnetic field on electrons. On the other hand, the distribution of high-energy electron impacts on the MCP is more uniform.

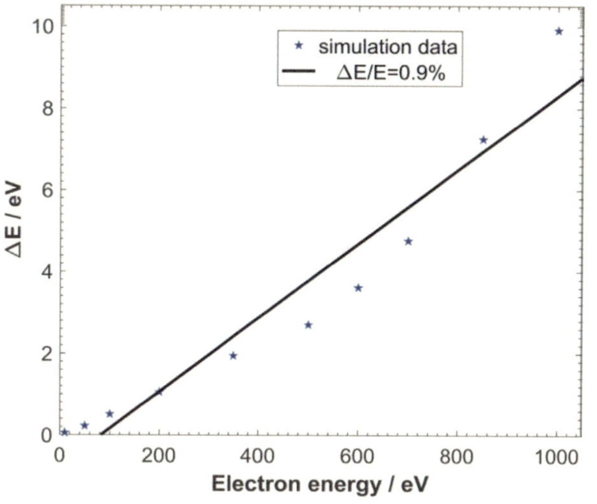


**Figure 19.**
*Calculated energy resolution of the magnetic-bottle time-of-flight spectrometer.*

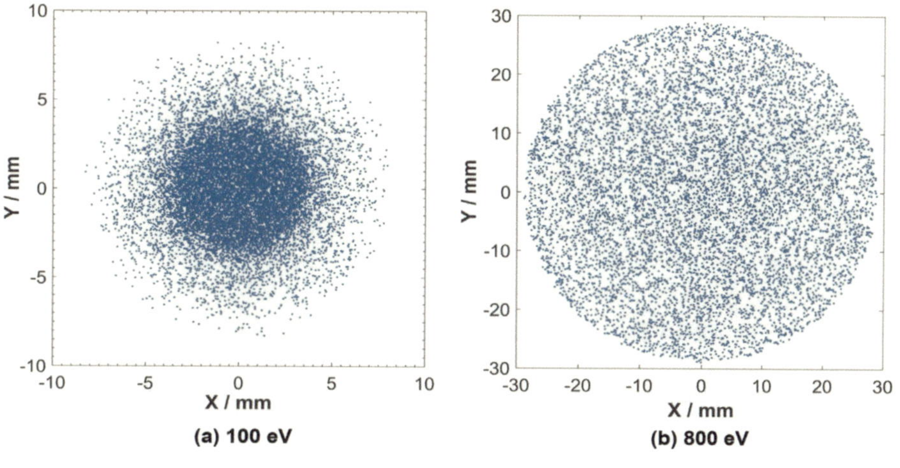


**Figure 20.**
*Electron trajectory tracking in magnetic field system: impact position distribution on the MCP of (a) 100 eV and (b) 800 eV electrons.*

## 4. Conclusions

In this chapter, we propose a novel physical model for analyzing the self-broadening dynamics of electron packets, the reduced thin electron disk model. It is found that the self-broadening process consists of two sequential steps: an early happening avalanche explosion and a subsequent slow quasi-linear spreading. The avalanche explosion occurs within such an extremely localized time and space range that the initial parameters of the electron packet have little effect on the broadening magnitude in this stage, while the initial parameters have complicated coupling effects on the quasi-linear spreading process. The avalanche broadening is identified as the dominating factor for the electron packet broadening process. Aimed at the primary requirement for constructing an adiabatic magnetic field used by a magnetic-bottle time-of-flight spectrometer, a practical scheme for a side pole-piece magnetic lens is proposed, and the high peak-valley ratio adiabatic magnetic field is constructed by the side pole-piece magnetic lens, the permanent magnet pole piece, and the solenoid. Finally, we demonstrate a design scheme for a magnetic-bottle electron energy spectrometer. The maximum detectable electron energy can reach 1000 eV, the full-range relative energy resolution is better than 1%, and the collection solid angle reaches 2π. This is an innovation with significant theoretical value, which will not only enrich the theory of high-precision detection and analysis techniques for electron energy spectra but also meet the urgent demand for high-performance electron momentum spectrometers in fields such as electron momentum spectroscopy, attosecond ultrafast spectroscopy, and photoelectron spectroscopy. At the same time, it can also serve as a direct reference for other charged particle analysis spectrometer technologies.

## Acknowledgements

This work was supported by the Natural Science Basic Research Program of Shaanxi Province (Program Nos. 2025JC-YBMS-101, 2022JZ-36).

## Author details

Yifan Kang[1,3,4], Hongtao Yang[1,3], Chao Wang[2*] and Wei Zhao[2,3*]

1 Xi'an Institute of Optics and Precision Mechanics, Chinese Academy of Sciences, Xi'an, China

2 Center for Attosecond Science and Technology, State Key Laboratory of Ultrafast Optical Science and Technology, Xi'an Institute of Optics and Precision Mechanics of Chinese Academy of Sciences, Xi'an, China

3 University of Chinese Academy of Sciences, Beijing, China

4 School of Science, Air Force Engineering University, Xi'an, China

*Address all correspondence to: igodwang@163.com; weiz@opt.ac.cn

IntechOpen

© 2025 The Author(s). Licensee IntechOpen. This chapter is distributed under the terms of the Creative Commons Attribution License (http://creativecommons.org/licenses/by/4.0), which permits unrestricted use, distribution, and reproduction in any medium, provided the original work is properly cited.

## References

[1] Peng P, Marceau C, Villeneuve DM. Attosecond imaging of molecules using high harmonic spectroscopy. Nature Reviews Physics. 2019;**1**:144-155. DOI: 10.1038/s42254-018-0015-1

[2] Rajput J, Severt T, Berry B, et al. Native frames: Disentangling sequential from concerted three-body fragmentation. Physical Review Letters. 2018;**120**(10):103001. DOI: 10.1103/PhysRevLett.120.103001

[3] He LX, Zhu XS, Cao W, Lan PF, Lu PX. Attosecond spectroscopy for filming the ultrafast movies of atoms, molecules and solids. Chinese Physics B. 2022;**31**(12):123301. DOI: 10.1088/1674-1056/aca6d2

[4] Ryabikin MY, Emelin MY, Strelkov VV. Attosecond electromagnetic pulses: Generation, measurement, and application. Attosecond metrology and spectroscopy (reviews of topical problems). Physics-Uspekhi. 2023;**66**(4):360-380. DOI: 10.3367/UFNe.2021.10.039078

[5] Geneaux R, Kaplan CJ, Yue L, et al. Attosecond time-domain measurement of core-level-exciton decay in magnesium oxide. Physical Review Letters. 2020;**124**(20):207401. DOI: 10.1103/PhysRevLett.124.207401

[6] Jia SK, Zhou JQ, Wang X, et al. Cold-target electron-ion-coincidence momentum-spectroscopy study of electron-impact single and double ionization of N2 and O2 molecules. Physical Review A. 2023;**107**(3):032819. DOI: 10.1103/PhysRevA.107.032819

[7] Wang SX, Zhu LF. Non-resonant inelastic X-ray scattering spectroscopy: A momentum probe to detect the electronic structures of atoms and molecules. Matter and Radiation at Extremes. 2020;**5**:054201. DOI: 10.1063/5.0011416

[8] Summers AM, Severino S, Reduzzi M, et al. Realizing attosecond core-level X-ray spectroscopy for the investigation of condensed matter systems. Ultrafast Science. 2023;**3**:Article 0004. DOI: 10.34133/ultrafastscience.0004

[9] Lindroth E, Calegari F, Young L, et al. Challenges and opportunities in attosecond and XFEL science. Nature Reviews Physics. 2019;**1**:107-111. DOI: 10.1038/s42254-019-0023-9

[10] Shi X, Liao CT, Tao ZS, et al. Attosecond light science and its application for probing quantum materials. Journal of Physics B: Atomic, Molecular and Optical Physics. 2020;**53**: 184008. DOI: 10.1088/1361-6455/aba2fb

[11] Garg M, Kern K. Attosecond coherent manipulation of electrons in tunneling microscopy. Science. 2020; **367**(6476):411-415. DOI: 10.1126/science.aaz1098

[12] Orfanos I, Makos I, Liontos I, et al. Attosecond pulsemetrology (review). APL Photonics. 2019;**4**(8):080901. DOI: 10.1063/1.5086773

[13] Gaunnitz T, Jain A, Pertot Y, et al. Streaking of 43-attosecond soft-X-ray packets generated by a passively CEP-stable mid-infrared driver. Optics Express. 2017;**25**(22):27506-27518. DOI: 10.1364/oe.25.027506

[14] Takahashi EJ, Lan P, Mücke OD, et al. Attosecond nonlinear optics using gigawatt-scale isolated attosecond pulses. Nature Communications. 2013;**4**: 2691. DOI: 10.1038/ncomms3691

[15] Pertot Y, Schmidt C, Matthews M, et al. Time-resolved X-ray absorption spectroscopy with a water-window high-harmonic source. Science. 2017; **355**(6322):264-267. DOI: 10.1126/science.aah6114

[16] Xu XR, Zhang YX, Zhang H, et al. Production of 100-TW single attosecond x-ray pulse. Optica. 2020;7(4):355-358. DOI: 10.1364/OPTICA.385147

[17] Li J, Lu J, Chew A, et al. Attosecond science based on high harmonic generation from gases and solids. Nature Communications. 2020;**11**:2748. DOI: 10.1038/s41467-020-16480-6

[18] Schoetz J, Wang Z, Pisanty E, et al. Perspective on petahertz electronics and attosecond nanoscopy. ACS Photonics. 2019;**6**:3057-3069. DOI: 10.1021/acsphotonics.9b01188

[19] Saha K, Pandey A, Banerjee SB, et al. Fragmentation kinematics of SF6 upon photo–excitation of S(2p) core shell and subsequent auger decays. Chemical Physics Letters. 2020;**739**:137038. DOI: 10.1016/j.cplett.2019.137038

[20] Ding X, Forbes R, Kübel M, et al. Threshold photo dissociation dynamics of $NO_2$ studied by time-resolved cold target recoil ion momentum spectroscopy. The Journal of Chemical Physics. 2019;**151**: 174301. DOI: 10.1063/1.5095430

[21] Weigold E, Hood ST, Teubner PJO. Energy and angular correlations of the scattered and ejected electrons in the electron-impact ionization of argon. Physical Review Letters. 1973;**30**(11): 475-478. DOI: 10.1103/PhysRevLett.30.475

[22] Cook JPD, McCarthy IE, Stelbovics AT, et al. Non-coplanar symmetric (e,2e) momentum profile measurements for helium: An accurate test of helium wavefunctions. Journal of Physics B: Atomic and Molecular Physics. 1984;**17**:2339-2352. DOI: 10.1088/0022-3700/17/11/025

[23] Ren XG, Ning CG, Deng JK, et al. (e, 2e) electron momentum spectrometer with high sensitivity and high resolution. The Review of Scientific Instruments. 2005;**76**(6):063103. DOI: 10.1063/1.1897668

[24] Shan X, Chen XJ, Zhou LX, et al. High resolution electron momentum spectroscopy of dichlorodifluoromethane: Unambiguous assignments of outer valence molecular orbitals. The Journal of Chemical Physics. 2006;**125**:154307. DOI: 10.1063/1.2358981

[25] Li J, Luo Z, Chen X, et al. Electron-impact ionization-excitation of the neon valence shell studied by high-resolution electron momentum spectroscopy. Physical Review A. 2015;**92**(3):032701. DOI: 10.1103/physreva.92.032701

[26] Pradhan E, Zeng T. The unified hamiltonian formalism of spin-orbit Jahn-teller and pseudo-Jahn-teller problems in all axial symmetries. Journal of Chemical Theory and Computation. 2023;**19**(21):7776-7786. DOI: 10.1021/acs.jctc.3c00782

[27] Xue H, Chen Y, Chen Q, et al. Deformation and spin-orbit splitting of hypernuclei in the Skyrme-Hartree-Fock approach. Physical Review C. 2023;**107**: 044317. DOI: 10.1103/PhysRevC.107.044317

[28] Li J, Liu Y. Hearing the heartbeat of atoms: Unveiling attosecond horizons. Ultrafast Science. 2023;**3**:Article 0049. DOI: 10.34133/ultrafastscience.0049

[29] Cao D, Xu W, Chen S, et al. Visualizing catalytic dynamics processes via synchrotron radiation multitechniques. Advanced Materials. 2023;**35**:2205346. DOI: 10.1002/adma.202205346

[30] Wang J, Driver T, Allum F, et al. Photon energy-resolved velocity map imaging from spectral domain ghost imaging. New Journal of Physics. 2023;**25**: 033017. DOI: 10.1088/1367-2630/acc201

[31] Meirer F, Weckhuysen BM. Spatial and temporal exploration of heterogeneous catalysts with synchrotron radiation(review). Nature Reviews Materials. 2018;**3**(9):324-340. DOI: 10.1038/s41578-018-0044-5

[32] Duris J, Li S, Driver T, et al. Tunable isolated attosecond X-ray packets with gigawatt peak power from a free-electron laser. Nature Photonics. 2019;**14**:30-36. DOI: 10.1038/s41566-019-0549-5

[33] Siol S, Mann J, Newman J, et al. Concepts for chemical state analysis at constant probing depth by lab-based XPS/HAXPES combining soft and hard X-ray sources. Surface and Interface Analysis. 2020;**52**(12):802-810. DOI: 10.1002/sia.6790

[34] Caciuffo R, Lander GH, Laan G. Synchrotron radiation techniques and their application to actinide materials. Reviews of Modern Physics. 2023;**95**: 015001. DOI: 10.1103/revmodphys.95.015001

[35] Ishikawa T. Accelerator-based X-ray sources: Synchrotron radiation, X-ray free electron lasers and beyond (review). Philosophical Transactions of the Royal Society A. 2019;**377**(2147):20180231. DOI: 10.1098/rsta.2018.0231

[36] Liu JC, Ignatova N, Kimberg V, et al. Time-resolved study of recoil-induced rotation by X-ray pump–X-ray probe spectroscopy. Physical Chemistry Chemical Physics. 2022;**24**:6627-6638. DOI: 10.1039/D1CP05000A

[37] Yamane H, Matsui F, Ueba T, et al. Acceptance-cone-tunable electron spectrometer for highly-efficient constant energy mapping. The Review of Scientific Instruments. 2019;**90**(9): 93102. DOI: 10.1063/1.5109453

[38] Hemmers O, Whitfield SB, Glans P, et al. High-resolution electron time-of-flight apparatus for the soft x-ray region. The Review of Scientific Instruments. 1998;**69**:3809. DOI: 10.1063/1.1149183

[39] Wu R, Ding B, Feng Y, et al. Scaling analysis parameters in 3D-focusing velocity map imaging. Measurement Science and Technology. 2023;**34**: 055502. DOI: 10.1088/1361-6501/acb72e

[40] Sparling C, Crane SW, Ireland L, et al. Velocity-map imaging of photoelectron circular dichroism in non-volatile molecules using a laser-based desorption source. Physical Chemistry Chemical Physics. 2023;**25**:6009. DOI: 10.1039/d2cp05880d

[41] Zhou L, Ni H, Jiang Z, et al. Ultrafast formation dynamics of D3+ from the light-driven bimolecular reaction of the D2–D2 dimer. Nature Chemistry. 2023; **15**:1229-1235. DOI: 10.1038/s41557-023-01230-0

[42] Tang T. Discrete space charge effect in charged particle beam and its calculation (in Chinese). Journal of Xi'an Jiaotong University. 1984;**18**(1):75-85

[43] Tao Z, Zhang H, Duxbury PM, et al. Space charge effects in ultrafast electron diffraction and imaging. Journal of Applied Physics. 2012;**111**(4): Article 044316. DOI: 10.1063/1.3685747

[44] Kruit P, Read FH. Magnetic field paralleliser for $2\pi$ electron-spectrometer and electron-image magnifier. Journal of Physics E: Scientific Instruments. 1983; **16**(3):313-324. DOI: 10.1088/0022-3735/16/4/016

Chapter 10

# Sustainable Electromagnetic Interference Shielding with Biomass Waste

*Venkatagiri Bharath and Sujatha Basa Reddy*

## Abstract

This chapter explores the sources of electromagnetic interference (EMI), its harmful effects on health and immunity, and its impact on interference and compatibility in power converters. While exploring shielding mechanisms of biomass-derived materials, showing their potential as eco-friendly alternatives to traditional EMI shielding solutions. With growing environmental concerns and increased EMI pollution from electronic devices, the need for eco-friendly, efficient shielding solutions is more pressing than ever. Biomass-derived composites, sourced from agricultural residues, forestry waste, and marine organisms, offer a unique combination of lightweight, biodegradable, and cost-effective properties. This chapter explores how biomass-derived materials naturally shield against electromagnetic interference through reflection, absorption, and multiple scattering, thanks to their unique structures and compositions. By comparing different biomass sources, processing methods, and hybrid composites, it highlights how their electromagnetic properties can be customized for various stealth and EMI shielding needs. This chapter emphasizes how sustainable material design can help tackle electromagnetic pollution while also being sustainable. By combining smart waste management with cutting-edge technology, biomass-derived composites stand out as great options for lightweight, high-performance EMI shielding in fields like aerospace, automotive, and wearable electronics.

**Keywords:** biomass-derived composites, electromagnetic interference (EMI) shielding, sustainable materials, stealth applications, environmental sustainability

## 1. Introduction

In an era driven by rapid technological advancements, electromagnetic fields permeate nearly every aspect of modern life. Electromagnetic waves (EMW) enable high-speed data transmission, wireless connectivity, and precision imaging. One of the most pressing challenges associated with EMW is electromagnetic interference (EMI). EMI is emerging as a major concern, as it impairs the functionality of modern electronic devices and poses challenges to their optimal performance. EMI presents a significant obstacle to the widespread adoption of 6G technology, potentially limiting its efficiency and reliability in next-generation communication systems. In ultra-fast wireless networks, interference causes delays and data loss, making real-time applications like remote surgeries and self-driving cars less reliable. In crowded urban areas, signals from nearby devices overlap, leading to unstable connections. In smart factories, EMI can disrupt communication between machines, causing errors and reducing efficiency. There are two approaches to addressing this issue. The first approach focuses on developing wireless communication technology to manage EMW propagation where the EMW are directionally transferred. In this approach, the signals should be transferred solely between the transmitter and receiver, not throughout the environment. While this seems promising, long-distance communication is a significant hurdle. The second approach is more practical as it uses materials that can attenuate the EMI. Basically, these materials counter interference in two ways:

a. By reflecting interference signals away.

b. By absorbing EM energy and converting it into heat instead of allowing the interference signal to bounce back.

As electronic devices become more integrated into daily life, EMI poses increasing risks, not just to technology but also to human safety. Medical implants like pacemakers and hospital monitoring systems can malfunction due to unintended electromagnetic signals, leading to serious health consequences. Similarly, EMI can disrupt navigation and control systems in aircraft and automobiles, posing safety hazards. Beyond immediate risks, concerns have been raised about prolonged exposure to high-intensity electromagnetic fields, with potential effects on neurological and cardiovascular health. To tackle this, metals and conductive polymers have traditionally been used as shielding materials. While effective, these materials come with drawbacks—they are heavy, prone to corrosion, expensive, and not environmentally friendly [1]. Because of these challenges, researchers are now exploring sustainable, lightweight, and cost-effective alternatives for EMI shielding. Biomass-derived carbon materials have emerged as promising candidates for EMI shielding due to their tunable electrical conductivity, dielectric properties, and eco-friendly nature.

This chapter begins by discussing the sources of EMI and its growing impact on modern electronic systems and human health. It then explores the mechanisms behind EMI shielding, focusing on how materials interact with EM waves through reflection, absorption, and multiple reflections. Next, the potential of biomass-derived materials as sustainable, eco-friendly alternatives for EMI attenuation is examined, highlighting their unique property and advantages over conventional shielding materials. Finally, the chapter delves into challenges and potential solutions with key insights into the development of sustainable EMI shielding solutions.

## 2. Electromagnetic interference and its impact

Understanding the impact of EMI requires a closer look at its origins, EMI originates from both natural and man-made sources, each contributing to the complex EM environment we inhabit, influencing not only electronic systems but also biological environments.

*DOI: http://dx.doi.org/10.5772/intechopen.1009943*

## 2.1 Sources of EMI

### *2.1.1 Natural sources*

Natural sources of EMI have always been present in our environment, affecting electronic devices and communication networks long before the emergence of man-made sources. Cosmic radiation, solar flares, and lightning are examples of natural EMI sources, as represented in **Figure 1**. Strong EM pulses produced by lightning strikes have the ability to disrupt sensitive electrical devices and communication networks over long distances. The EM field fluctuates as a result of high-energy particles released by solar flares and geomagnetic storms that originate from the sun and interact with the Earth's magnetosphere. Power grids, GPS systems, and satellite communications can all be affected by these solar-induced disruptions. Background noise from cosmic radiation from deep space affects high-frequency communication systems and radio telescopes.

### *2.1.2 Man-made sources*

Man-made sources of EMI are increasingly prevalent due to the rapid advancement of technology and the growing use of electronic devices. These sources can be broadly categorized into automotive and industrial mobility, electromagnetic wave transmission, power generation and broadcasting, high-power machinery and tools, and common household devices, as represented in **Figure 1**. EM waves released by communication equipment including satellite links, mobile transmitters, radar installations, and broadcast stations can disrupt other devices using the same frequencies. Due to high current flow and switching transients, industrial equipment such as electric motors, power generators, and welding machines produces significant EMI. Commonplace household appliances like microwaves, PCs, and fluorescent lights can cause EMI that can affect other electronic devices in the vicinity. EMI is produced by ignition systems, electronic control units, and power converters in transportation systems, such as cars and airplanes.

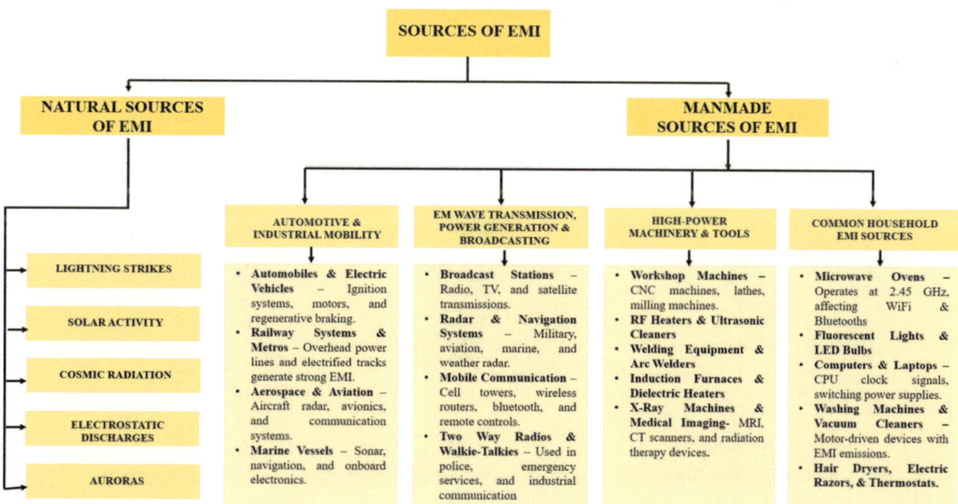


**Figure 1.**
*Sources of EMI.*

### 2.2 Effects of EMI on human health

The potential effects of EM waves on human health have been widely studied, particularly in the low-frequency and microwave ranges. While high-intensity RF radiation is known to cause thermal effects, the debate continues regarding non-thermal biological effects.

High-frequency electromagnetic fields primarily affect tissues through heating effects. However, some studies suggest that low-intensity RF fields may influence cellular functions, although conclusive evidence remains elusive. Neurological effects of RF radiation have also been explored, with early research, particularly from the Soviet Union, investigating its impact on nerve function, behavior, and brain chemistry. Experimental studies have indicated that microwave exposure at high specific absorption rates (SAR > 5 W/kg) can alter neural activity in isolated nerve preparations. **Figure 2** depicts the depth of penetration of cell phone radiation in children and adults [2]. Nevertheless, no consistent effects on brain function have been observed at lower exposure levels. In addition to neurological implications, cardiovascular and physiological effects have been examined. Studies on extremely low-frequency fields, such as 50 Hz fields comparable to those from high-voltage power lines, suggest minimal impact on human physiology. While minor fluctuations in heart rate and reaction times have been reported, these variations fall within normal biological ranges. Prolonged exposure to high-intensity RF radiation has also been linked to ophthalmic concerns, particularly the development of cataracts. This risk is more pronounced in the Giga frequency range, where power densities exceeding 100 mW/cm$^2$ for extended periods can damage the eye lens. However, public exposure levels from communication systems are typically far below these harmful thresholds.

### 2.3 Electromagnetic interference and compatibility in power converters

Effective power converter design is essential for minimizing power losses and EMI noise. Engineers can improve efficiency, increase system dependability, and reduce filtering costs by addressing EMI concerns early in the design process. A well-designed power converter topology maximizes electromagnetic compatibility (EMC) while retaining high performance. Several key factors influence EMI in power converters. One of the most significant contributors is switching transients, which occur due to the rapid change in voltage and current during switching events. These

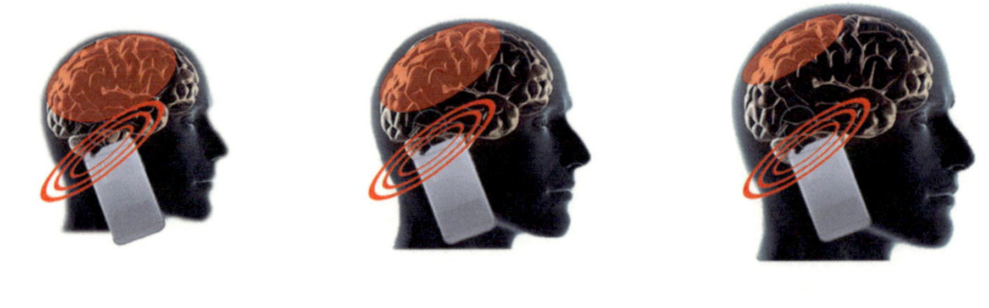


**Figure 2.**
*Radiation exposure and depth of penetration.*

transients can generate high-frequency noise and EMI. Selecting appropriate switching components and employing advanced modulation techniques can effectively mitigate these effects. The choice of pulse-width modulation (PWM) technique and converter topology also significantly impacts the EMI profile of power converters. Different PWM schemes generate varying harmonic content and switching noise. Careful selection of modulation methods can help control harmonics and reduce unwanted emissions, ensuring compliance with EMC standards. Another critical aspect of EMI management is filter design. Passive EMI filters are commonly used to suppress conducted and radiated noise. However, the size and cost of these filters depend on several factors, including switching frequency, ripple voltage, and overall system losses. Proper selection of switching parameters, such as frequency and duty cycle, can minimize the required filter size and complexity, leading to a more cost-effective design.

### 2.4 Effects of electromagnetic fields on the immune system

The potential impact of electromagnetic fields (EMFs) on the immune system has been an area of active research, with studies exploring both radiofrequency (RF) fields and extremely low-frequency (ELF) fields. Research findings have varied significantly, reflecting the complexity of the immune system's response to electromagnetic exposure.

#### *2.4.1 RF radiation and immune cell response*

Numerous in vitro (lab-based) and in vivo (animal-based) studies have investigated how RF fields affect immune system cells. Some studies suggest that exposure to RF fields may influence lymphoblast transformation and immune responses to mitogens, which are substances that stimulate cell division. However, the observed effects have been inconsistent across different research settings, underscoring the variability in experimental conditions and biological responses. A specific absorption rate (SAR) threshold greater than 4 W/kg has been linked to changes in lymphocyte (white blood cell) responses to mitogen stimulation under pulsed and continuous-wave (CW) microwave exposure. In addition, thermogenic exposure—where RF radiation induces tissue heating—has been associated with reduced natural killer (NK) cell activity and increased macrophage activation. This phenomenon is thought to result from elevated steroid hormone release into the bloodstream due to thermal stress. Interestingly, a long-term study investigating non-thermal pulsed 2.45 GHz microwaves (SAR ~0.4 W/kg) on rats over two years revealed no significant long-term changes in lymphocyte counts or overall immune function. This finding highlights the complexity of RF radiation effects, suggesting that the impact on the immune system may depend on exposure duration, frequency, and power levels.

#### *2.4.2 Effects of extremely low-frequency fields on immunity*

Research into ELF electric and magnetic fields has generally shown minimal impact on immune function in laboratory animals exposed under controlled conditions. However, some in vitro studies have reported that pulsed magnetic fields and 50 Hz amplitude-modulated RF fields may reduce mitogen responses and target-cell toxicity. This is believed to be due to high current densities generated in the cell suspensions, which could alter cellular signaling pathways. One study simulated exposure conditions near high-voltage power lines, using 60 Hz electric and magnetic fields, and observed no significant effects on the immune functions of human and canine lymphocytes. This was true for both healthy donors and those challenged with specific antigens, indicating that typical environmental exposure levels may not pose a significant risk to immune health.

## 3. Mechanisms of EMI shielding in biomass-derived materials

Shielding effectiveness (SE) is a critical measure in evaluating a material's protective capabilities, and assesses its ability to block EMIs. EMIs are comprised of electric and magnetic components propagating at the same frequency in perpendicular orientation to each other. Materials with high conductivity effectively attenuate the electric field and enhance magnetic permeability, thereby influencing both fields within an electromagnetic environment, suppressing one invariably affects the other.

### 3.1 Shielding mechanisms

As SE results from absorption ($SE_A$), reflection ($SE_R$), and multiple reflections ($SE_{MR}$). According to Schelkunoff's theory, the comprehensive shielding effectiveness ($SE_T$) represents the cumulative impact of these phenomena, as shown in eq. (1) [3]:

$$SE_T = SE_A + SE_R + SE_{MR} \tag{1}$$

According to Schelkunoff's theory, when the absorption loss exceeds 10 dB, the contribution from multiple internal reflections ($SE_{MR}$) becomes negligible. In such cases, the total shielding effectiveness is simplified, as shown in eq. (2) [3]. EMI shielding mechanism is depicted in **Figure 3**:

$$SE_T = SE_A + SE_R \tag{2}$$

Biomass-derived carbon nanohybrids exhibit high electrical conductivity due to their graphitized carbon structures. This high conductivity generates robust surface currents that reflect incident electromagnetic waves, contributing to SER. Additionally, the presence of conductive fillers such as graphene nanoplatelets (GNPs) or carbon black enhances reflection through impedance mismatch at the material-air interface. Eq. (3) [3] relates power (P), electric field (E), and magnetic field (H), where PT and PI denote the transmitted and incident wave power, respectively; the expression for shielding effectiveness (SE) can be formulated as per eq. (3)

$$SE(dB) = -10 log\, log\left(\frac{P_t}{P_i}\right) = -20 log\, log\left(\frac{H_t}{H_i}\right) = -20 log\, log\left(\frac{E_t}{E_i}\right) \tag{3}$$

These nanocomposites effectively absorb EM waves due to multiple polarization mechanisms, including dipole polarization and interfacial polarization. The presence of heterogeneous interfaces between the biomass-derived carbon and polymer matrix enhances charge carrier mobility, leading to increased dielectric losses. The hierarchical porous structure of these nanocomposites facilitates multiple internal reflections, extending the EM wave propagation path and enhancing energy dissipation through eddy

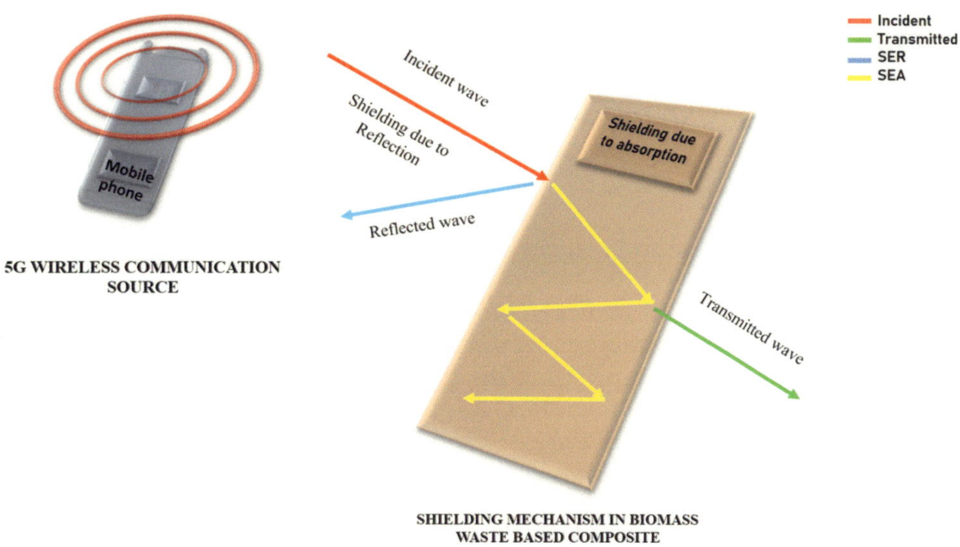


**Figure 3.**
*Mechanism of EMI shielding.*

currents. The SEA is a result of the formation of electric and magnetic dipoles that absorb EM waves. Materials with a high ability to store electrical energy (dielectric constant) and respond to magnetic fields (magnetic permeability) are effective for EMI shielding. As the EM waves travel through the material, their magnetic fields are intercepted. Adequate absorption takes place by attenuating the EM wave through eddy currents (currents induced by the high-frequency EM wave that oppose external magnetic fields). Materials with high conductivity induce more robust eddy currents. The porous morphology and layered architecture of biomass-derived nanohybrids encourage multiple internal reflections, increasing the effective path length of EM waves within the material. This mechanism enhances energy dissipation, particularly in low-frequency ranges. However, when SEA exceeds 10 dB, SEMR becomes negligible due to substantial attenuation within the material layers. The $SE_T$, $SE_R$ and $SE_A$ in terms of S parameters are expressed in eq. (4)–(6) [3]. The effectiveness of EMI shielding is strongly influenced by the frequency of the incident EM field. The two primary shielding mechanisms SER and SEA respond differently to frequency changes:

$$SE_T = 10\,log\,log\left(\frac{1}{|S_{12}|^2}\right) \tag{4}$$

$$SE_R = 10\,log\,log\left(\frac{1}{1-|S_{11}|^2}\right) \tag{5}$$

$$SE_A = 10\,log\,log\left(\frac{1-|S_{11}|^2}{|S_{12}|^2}\right) \tag{6}$$

S11, S21 and S12 represent the reflection, reverse transmission, and forward transmission coefficients, respectively. These scattering parameters can also be utilized to find the dielectric properties utilizing Nicholson Rose Weir (NRW), Short Circuit Line (SCL), New non-iterative (NNI), and National Institute of Standards and Technology (NIST) iterative conversion techniques. A comparison of various conversion techniques is given in **Table 1**.

### 3.2 Influence of dielectric and magnetic properties on EMI shielding

The EMI shielding performance of biomass waste-derived nanocomposites is significantly influenced by their dielectric and magnetic properties. These properties are dictated by the composition, morphology, and interfacial interactions within the nanocomposite matrix. Biomass-derived carbon nanohybrids exhibit high permittivity due to the presence of conductive networks formed by interconnected carbon nanoparticles. These conductive pathways enable interfacial polarization, where charge carriers accumulate at the heterogeneous interfaces between the carbon fillers and polymer matrix. This polarization enhances dielectric losses, resulting in effective EM wave absorption. Additionally, the hierarchical porous structure facilitates dipole polarization, further increasing the attenuation of high-frequency electromagnetic waves. The incorporation of magnetic nanoparticles, such as ferrites, into biomass-derived nanocomposites introduces frequency-dependent permeability, enabling tunable magnetic loss mechanisms. These magnetic fillers absorb the magnetic component of incident EM waves through natural resonance (alignment of magnetic dipoles with the external field) and eddy current loss (induced circulating currents opposing the incident field). This combination effectively reduces reflection and enhances absorption, contributing to high absorption-based shielding effectiveness. The synergistic effect of dielectric and magnetic losses contributes to high absorption-based shielding effectiveness. High electrical conductivity is achieved through the graphitization of biomass-derived carbon, which forms a continuous network of conductive paths. This structure supports robust eddy currents that oppose incident EM fields, enhancing reflection-based shielding. The addition of conductive fillers, such as graphene nanoplatelets and carbon black, further boosts conductivity, leading to enhanced SER. However, an optimal balance between conductivity and permittivity is essential to minimize secondary reflections. For instance, the combination of magnetic nanoparticles with conductive carbon fillers leads to impedance matching, which minimizes reflection and maximizes absorption. The hierarchical architecture and porous morphology enable multiple scattering and internal reflections, thereby enhancing the overall shielding of EMI.

| Conversion techniques | S parameters | Dielectric properties |
|---|---|---|
| NRW | S11, S21, S12, and S22 | $\varepsilon_r$ and $\mu_r$ |
| NIST | S11, S21, S12, and S22 | $\varepsilon_r$ and $\mu_r = 1$ |
| SCL | S11 | $\varepsilon_r$ and $\mu_r = 1$ |
| NNI | S11, S21, S12, and S22 | $\varepsilon_r$ |

**Table 1.**
*A comparison of various conversion techniques.*

## 4. Biomass materials for EMI shielding

The utilization of biomass waste in electromagnetic shielding applications aligns with the principles of sustainable development, the primary motivation for adopting biomass-based shielding materials stems from their:

a. Abundance and renewability: Biomass waste, such as coconut shells, rice husks, wood chips, and agricultural residues, is widely available and can be transformed into carbon-rich conductive materials.

b. Lightweight nature: Unlike metals, biomass-derived shielding materials possess low density, making them ideal for aerospace, automotive, and wearable electronic applications.

c. Eco-friendliness: The conversion of biomass waste into shielding materials reduces environmental impact, promoting a circular economy and sustainable waste management.

d. Cost-effectiveness: As compared to high-cost synthetic conductive polymers or graphene-based composites, biomass-derived carbon materials offer a low-cost alternative with comparable shielding efficiency.

Using fruit peels not only provides a sustainable option, but it also helps to manage organic waste, which would otherwise pollute the environment. it was estimated that roughly 150 billion metric tonnes of agricultural biomass waste are produced worldwide each year [4]. Improper disposal and inefficient use of this waste contribute to major environmental issues such as water and soil contamination, deterioration of air quality, climate change, and depletion of natural resources [5]. The synergy between effective waste management and advanced technological applications positions fruit-based composites as a promising solution for environmentally friendly EMI shielding materials.

### 4.1 Banana leaf and pineapple fiber-based carbon composites

In the study by Raji et al. [6], deals with banana leaf-derived carbon composite. The banana leaf is grinded and dispersed in sucrose solution, after uniform mixing they are subjected to filter pressing and freeze drying, which results in a banana leaf fiber sucrose (BL/FS) composite. This BL/FS composite is subjected to carbonization at 900°C to produce foams (sucrose acts as a fiber to hold the BL carbonized fibers). It is reported the electrical conductivity of BL/FS composite increases (3.7–8.7 S $m^{-1}$), with increase in concentration of sucrose solution (300–700 g $L^{-1}$) reaching the maximum SET of 77.5 dB in X band. In the study by Balaji Krishnabharathi et al. [7], a composite material reinforced with 40 vol% pineapple fiber (PAF) and 4% biocarbon, which were derived from abaca bract and pineapple fiber through pyrolysis was created using a rigid vinyl ester matrix (PAF-BC/VEM). It is reported that this composite achieved EMI SE of 11.18 dB at 8 GHz, 30.53 dB at 12 GHz, 54.16 dB at 16 GHz and 65.51 dB at 20 GHz.

### 4.2 Bamboo fiber and wood-derived-based composites

In the study by Zhang et al. [8], an interesting case of 'multi-absorption-reflection-reabsorption' of EMW is presented. Bamboo fibers (BF) and biodegradable polylactic acid (PLA) were used a raw matrix to fabricate alternative bamboo fibers coated with copper(conductive) and iron(magnetic) multilayer film composite((Fe/Cu)@BF/PLA) using solution casting and hot pressing method. Alternative odd multilayer stacking from 3 to 9 was studied, when subjected to shielding tests in X band, 3 layers had 45.02 dB and 9 layers had 80.2 dB SET. A layer optimization technique (varying the thickness of the Cu and Fe layers in a controlled manner) could lead to more efficient multi-absorption and reflection, optimizing performance while maintaining a thinner composite. The 'multi-absorption-reflection-reabsorption mechanism demonstrated by the (Fe/Cu)@ BF/PLA composite highlights the unique advantage of using alternating conductive and magnetic layers for EMI suppression. This layering allows for superior SE performance at reduced material thickness, emphasizing the potential of bamboo fiber composites as lightweight, high-performance EMI shielding solutions. In the study by Mayin et al. [9], natural poplar (light colored/low-density hardwood timber) was subjected to carbonization at $700^{0}$ C. This wood-derived carbon was used as a matrix which was impregnated with nickel-magnetite @ graphene (Ni-$Fe_3O_4$@GNP) with different impregnation time (3, 5, 6, and 9 hours). It is reported that the 7-hour impregnated CW/ (Ni-$Fe_3O_4$@GNP) composite attenuated 99.999996% (84.25 dB) of EM waves throughout X band.

### 4.3 Chitin, chitosan, and cellulose-based aerogels

In the study by S Wang et al. [10], A CTS-based porous carbon/cobalt aerogel (CTS-C/Co aerogel) was fabricated via freeze-drying and carbonization. In-situ grown cobalt particles enhanced EMI absorption, achieving a – 75.8 dB absorption, −256.9 dB $mm^{-1}$ reflection loss, and an 8.53 GHz effective absorption bandwidth at 10% filler content. In the study by Zhang X et al. [11], Chitin (CTN)-based nanofiber aerogels were developed by cross-linking chitin and aramid nanofibers (CTN-ANF) using a synchronous deprotonation-protonation method. The resulting CTN-ANF aerogels were further impregnated with Ti3C2Tx MXene to form conductive aerogels (CTN-ANF aerogel@Ti3C2Tx MXene), which demonstrated EMI SE of 75 dB in X band frequencies. In the study by Mengxin et al. [12], flexible film composites made from varying ratio of MXene nanosheets, liquid metal droplets and bacterial cellulose ($Ti_3C_2T_x$/LM/BTC) were fabricated through vacuum filtration and hot pressing technique. The film with 90 wt. % MXene, 5 wt. % BC and 5 wt. % LM exhibited the highest SE of 61.84 dB in X band. The study also reports the films having excellent electric conductivity (2867.9 S $m^{-1}$) and thermal conductivity (10.44 W $m^{-1}$ $K^{-1}$). In the study by Wang et al. [13], a conductive CTF embedded with in-situ reduced silver nanoparticles (Ag-CTF) is reported to offer high conductivity and water proof properties, with excellent SET of 102.32 dB in the X band and 107.62 dB in the K band. In the study by Saraswathy R et al. [14], carbon composite grids (CCGs) created from castoff cotton cloth by impregnating with a sucrose and phenol formaldehyde (PF) solution, followed by curing and carbonization at 900°C, the CCGs is reported to have should an EMI SE ranging from 34.8 to 52.4 dB in X band frequencies which is reflection dominant. In the study by Dai et al. [15], MWCNTs were added as functional additives to an eco-friendly CNF matrix for the synthesis of CNF/ MWCNTs films in various ratios. It is reported the film with 1:4 (CNF: MWCNTs) had an excellent conductivity of $1.1 \times 10^3$ S/m and had the highest shielding of 55.8 dB in the X band, as represented in **Table 2**.

| Composite | Biomaterial | SE value (dB) | Frequency (GHz) | Frequency band | References |
|---|---|---|---|---|---|
| BL/FS | Banana leaf, fiber sucrose | 77.5 | 8–12 | X | [6] |
| CW/(Ni-$Fe_3O_4$@ GNP) | Natural Poplar (timber wood) | 84.25 | 8–12 | X | [9] |
| CTS-C/Co aerogel | Chitosan | 75.8 (SEA) | 8.53 | X | [10] |
| CTN-ANF aerogel@Ti3C2Tx MXene | Chitin | 75 | 8–12 | X | [11] |
| Ti3C2Tx/LM/BTC | Bacterial cellulose | 61.84 | 8–12 | X | [12] |
| (Fe/Cu)@BF/PLA | BF, PLA | 80.2 | 8–12 | X | [8] |
| Ag-CTF | Cotton fabric | 102.32,107.62 | 8–12,18–27 | X, K | [13] |
| Carbon composite grids (CCGs) | Castoff cotton cloth, sucrose | 52.4 | 8–12 | X | [14] |
| CNF/MWCNTs | CNF | 55.8 | 8–12 | X | [15] |
| PAF-BC/VEM | Pineapple fiber, abaca fruit | 11.18, 54.16, 65.51 | 8–20 | X, Ku, K | [7] |

**Table 2.**
*A comparison of SE of various biomass-based composites.*

## 5. Challenges and future perspectives

Currently, there are complexities in producing biomass-derived composites at an industrial scale, including challenges in ensuring consistent quality and performance. Long-term performance and reliability could be affected by these materials' durability under harsh environmental conditions, such as high humidity, temperature fluctuations, and exposure to chemicals. Although biomass-derived composites offer eco-friendly benefits, a comprehensive assessment of their environmental impact is necessary to accurately compare their ecological footprint with that of conventional materials. Cost–benefit analyses are also needed to evaluate whether scaling these materials for industrial use is economically viable, especially considering high manufacturing costs and the need for advanced processing technologies. To advance the field, future research should focus on several targeted areas:

- Development of hybrid composites: Integrating biomass-derived materials with high-performance additives, such as magnetic nanoparticles, conductive fillers, and carbon-based nanomaterials, could enhance both mechanical properties and EMI shielding effectiveness. These hybrid composites could provide tailored properties for specific applications, from consumer electronics to aerospace.

- Multifunctional composite design: Research should explore composites with additional functional attributes, such as flame resistance, self-cleaning surfaces, or UV resistance. These features would expand the practical applications of biomass-based composites, improving their appeal and usability in sectors requiring both functional and esthetic qualities, like construction and automotive interiors. Integrating smart sensing technologies into these composites could also enable real-time monitoring of structural health and performance, enhancing safety and reliability.

- Improved manufacturing methods: Future work should investigate advanced manufacturing techniques, such as 3D printing, electro spinning, and solvent casting, to improve procrmity. Process optimization is essential to overcome technical difficulties in scaling and commercializing biomass-derived composites. Research into automation and precision-controlled manufacturing methods could also reduce costs and improve consistency.

- Testing under diverse environmental conditions: To ensure durability and reliability, composites should be tested under a range of environmental conditions, including prolonged humidity exposure, high temperatures, UV radiation, and mechanical stress. This research could inform design improvements that make these materials more resilient for outdoor or extreme environments. Accelerated aging tests and in-situ performance monitoring could provide valuable insights into long-term durability and maintenance requirements.

- Collaboration for commercialization: Partnerships between industrial stakeholders and academic institutions are crucial to bridging the gap from laboratory-scale research to commercial-scale production. Collaboration can expedite the development of commercially viable production methods, address regulatory hurdles, and improve supply chain integration for biomass-derived materials. Establishing industry standards and certifications will be critical for gaining market acceptance, especially in highly regulated sectors such as automotive, aerospace, and construction.

- Sustainability and life-cycle analysis: To validate the environmental benefits of biomass-derived composites, future studies should conduct life-cycle assessments that encompass raw material sourcing, production, usage, and disposal. This analysis will help determine the true ecological impact of these materials compared to conventional options and guide sustainable design choices. Exploring circular economy models, such as recycling and biodegradability, will enhance the sustainability profile of these composites.

## 6. Conclusion

This chapter explores how biomass-derived materials can block EMI, showing they are a promising, eco-friendly alternative to traditional shielding methods. Key takeaways include: Biomass-derived materials shield against EMI through reflection, absorption, and multiple scattering, thanks to their unique microstructures and dielectric properties. Sources such as agricultural residues, forestry waste, and marine biomass provide renewable, lightweight, and biodegradable options for shielding composites. Incorporating conductive fillers, such as carbon-based nanomaterials and magnetic particles, enhances the shielding performance of these materials by improving their electrical conductivity and magnetic responsiveness. Additionally, techniques like pyrolysis and chemical modification enable precise control over their electromagnetic characteristics, allowing them to be tailored for different frequencies and a wide range of applications. Looking forward, biomass-derived materials have great potential for EMI shielding in next-gen electronics and stealth tech. They're sustainable, affordable, and can be tailored for specific electromagnetic needs, making them perfect for eco-friendly shielding. As research continues into new biomass sources, better processing techniques, and advanced hybrid composites, these materials are set to become even more versatile, opening up exciting possibilities for green solutions in electronics and stealth applications.

## Conflict of interest

The authors declare no conflict of interest.

## Author details

Venkatagiri Bharath[1,2]* and Sujatha Basa Reddy[2]

1 HKBK College of Engineering, Bangalore, India

2 M S Ramaiah Institute of Technology, Bangalore, India

*Address all correspondence to: bharathv.work@gmail.com

IntechOpen

© 2025 The Author(s). Licensee IntechOpen. This chapter is distributed under the terms of the Creative Commons Attribution License (http://creativecommons.org/licenses/by/4.0), which permits unrestricted use, distribution, and reproduction in any medium, provided the original work is properly cited.

## References

[1] Reddy SB, Bharath V, Naik DR. Structural, morphology and electromagnetic interference shielding studies of multifunctional hybrids of low-density polyethene with graphene, carbon nanotube and magnetite. Polymer-Plastics Technology and Materials. 2025;**64**(3):364-381

[2] Gandhi OP, Morgan LL, De Salles AA, Han YY, Herberman RB, Davis DL. Exposure limits: The underestimation of absorbed cell phone radiation, especially in children. Electromagnetic Biology and Medicine. 2012;**31**(1):34-51

[3] Reddy SB, Bharath V. A comprehensive review on the characterization of nanomaterials for the suppression of electromagnetic interference across diverse frequency bands. Polymer-Plastics Technology and Materials. 2024;**63**(14):1917-1958

[4] Asim N, Emdadi Z, Mohammad M, Yarmo MA, Sopian K. Agricultural solid wastes for green desiccant applications: An overview of research achievements, opportunities and perspectives. Journal of Cleaner Production. 2015;**91**:26-35

[5] Wang B, Dong F, Chen M, Zhu J, Tan J, Fu X, et al. Advances in recycling and utilization of agricultural wastes in China: Based on environmental risk, crucial pathways, influencing factors, policy mechanism. Procedia Environmental Sciences. 2016;**31**:12-17

[6] Raji S, Sharma GK, Aranya BR, Prabhakaran K. Carbon composite foams from the wasted banana leaf for EMI shielding and thermal insulation. Carbon. 2023;**213**:118259

[7] Balajikrishnabharathi A, Jayabalakrishnan D. Characterization of mechanical, dielectric, EMI shielding properties of abaca bract biocarbon and pineapple fiber reinforced rigid vinyl ester composite. Polymer Bulletin. 2024;**81**(13):12267-12285

[8] Zhang Q, Tang X, Zhao Q, Chen X, Wang K, Zhang Q, et al. Millefeuille-Inspired Biomass Alternate Multilayer Composite for Excellent Absorption-Dominated, Broadband EMI Shielding and Joule Heating. Composites Part B: Engineering; 2024. p. 111620

[9] Dai M, Pan Y, Guo Q, Yang D, Qiu F, Hu S, et al. Nickel-$Fe_3O_4$@graphene coated lightweight carbonized wood for efficient electromagnetic interference shielding. Diamond and Related Materials. 2024;**146**:111242

[10] Wang S, Zhang X, Tang Y, Hao S, Zheng S, Qiao J, et al. Facile fabrication of biomass chitosan-derived magnetic carbon aerogels as multifunctional and high-efficiency electromagnetic wave absorption materials. Carbon. 2024;**216**:118528

[11] Zhang X, Qian K, Fang J, Thaiboonrod S, Miao M, Feng X. Synchronous deprotonation–protonation for mechanically robust chitin/aramid nanofibers conductive aerogel with excellent pressure sensing, thermal management, and electromagnetic interference shielding. Nano Research. 2024;**17**(3):2038-2049

[12] Liu M, Zhang H, Zhang Z, Zhang K, Chen Z, Pan L, et al. Jujube-cake inspired lightweight and flexible MXene/liquid metal/bacterial cellulose electromagnetic interference shielding film with excellent joule heating and thermal conductivity. Composites Science and Technology. 2024;**255**:110722

[13] Wang Z, Xing D, Yin R, Zhan P, Liu C, Shen C, et al. Breathable and waterproof conductive cotton fabric pressure sensor with distinguished electrothermal and electromagnetic interference shielding performances. Applied Materials Today. 2024;**38**:102256

[14] Saraswathy R, Kavya MR, Prabhakaran K. Upcycling of used cotton cloth to robust fire-resistant carbon composite grids with excellent EMI shielding. ACS Applied Engineering Materials. 2024;**2**(6):1624-1639

[15] Dai M, Zhou D, Ren H, Deng S, Jiang Y, Sun J, et al. Cellulose-based multifunctional film for electromagnetic interference shielding, joule heating, and flame retardancy. Diamond and Related Materials. 2024;**143**:110883